Young Men in Harm's Way

Young Men in Harm's Way

SSG Arnold Krause

Deeds Publishing | Athens

Copyright © 2020 — Arnold Krause

ALL RIGHTS RESERVED—No part of this book may be reproduced in any form or by any electronic or mechanical means, including information storage and retrieval systems, without permission in writing from the authors, except by a reviewer who may quote brief passages in a review.

Published by Deeds Publishing in Athens, GA
www.deedspublishing.com

Printed in The United States of America

Cover design by Mark Babcock.

HARDCOVER ISBN 978-1-950794-04-1
PAPERBACK ISBN 978-1-950794-03-4

Books are available in quantity for promotional or premium use. For information, email info@deedspublishing.com.

First Edition, 2020

10 9 8 7 6 5 4 3 2 1

2nd Battalion, 12th Infantry Regiment

To the memory of the 328 officers and men who lost their lives during the four and half years the unit was called into service in Vietnam, you shall never be forgotten. And to the veterans with whom I proudly fought side by side with, you will always be my brothers.

"Banded by War, Brothers Forever"

The 2/12th Infantry served with honor and distinction, having been credited with 11 campaign ribbons, Presidential Unit Citation, 3 Republic of Vietnam Cross of Gallantry with Palm Unit Citations (1966-67, 1967-68, 1968-70), Republic of Vietnam Civil Action Honor Medal, 1st Class (1967-70)

Medal of Honor awarded posthumously to SP4 Donald Ward Evans, Jr., HHC, Medic, attached to A Co (January 27, 1967) 13 Distinguished Service Crosses, 93 Silver Stars

12th Infantry "Warrior" Regiment Motto: "Having Been Led By Love of Country"

CONTENTS

Preface	xi
History of the Unit	xvii
1. Operation Wilderness	1
2. Drafted—Civilian life ends	19
3. Leading up to my arrival in Vietnam	31
4. Dau Tieng and Camp Rainier	37
5. Michelin Rubber Plantation	51
6. Radio Telephone Traffic and language	65
7. Eagle Flight	77
8. Village Life	99
9. Trang Bang and Fullback 6	107
10. Tactics Learned and Used	129
11. Monsoon in the Rubber Plantations	137
12. July in Hoc Mon	149
Photographs	157
13. Life in a FSB	173
14. Medcaps, traffic control, cordon and searches and AP's	191
15. South Vietnam's military units	199
16. September—FSB Stuart in Trang Bang	203

17. Air Ambulance — How its call sign came to be	237
18. Fire Support Base Pershing	241
19. December, Our Worst Month	255
20. January, the Year of the Rooster and TET again	275
21. Patrol Base Granite — and home	293
22. Destination U.S.A.	305
Service record	315
Epilogue	317
Glossary	323
Roster of Charlie Company Enlisted (Drafted) Men	343
The Top 100 tunes — 1968	359
Half Truths and lies	363
About the Author	365

PREFACE

I originally started portions of this book, more like a diary, trying to create a *timeline* of the events of my military service. This effort served two purposes; one to satisfy my own quest to preserve on paper what I had achieved and experienced as a young soldier and the second reason, was the remote possibility that someone in my family, present and future, might have an interest in my military service.

My memory was very good for some periods of time and totally lacking or void for others, an experience shared by almost everyone I served with in Vietnam. In order to help myself along in this process of recollection, I resorted to searching the internet and discovered a number of military websites that had some information relating to my service in Vietnam, but that wasn't enough. In the end I started applying for records of my unit, the 2nd Battalion, 12th Infantry Regiment stored at College Park, Maryland at NARA, the National Archives and Records Administration. They are responsible for the storage of all military records of certain categories, depending on the statute of limitations imposed by the government. Several examples were the daily operational journals kept by battalions, LL's or *Lessons Learned* (sharing what to do or not do in various situations and usually written by division staff officers), operational reports etc. I had to keep my records request limited to small units, generally a month at a time since it might produce several hundred pages at .75 cents a copy. It was expensive, but in the overall view of what I wanted to accomplish, worth the price.

I was most interested in my unit's daily battalion journals. These were written by officers manning the TOC, tactical operations center, and were hour by hour journal entries of what was happening by every company or platoon in the battalion and where they were located and what they were doing twenty-four hours a day, There were also the LL's (Lessons Learned), quarterly reports written by division staff (summary of major events, statistical information by unit, and to some extent, interviews of major battles or conflicts), and lastly descriptions of the major operations conducted in the early years of the war. To get copies of personnel records for individuals and for company rosters of the men who served in the unit I was assigned to, it required more forms to be filled out and sent to St. Louis, Missouri, where the NPRC, National Personnel Records Center resides. Most of these records were free to obtain. The largest problem with St. Louis is time. They receive over 5,000 requests a day for information, so it takes months and months to receive a response once they logged in my request form, SF-180.

Armed with this information, and after conducting interviews with fellow veterans, I have been able to reconstruct the entire movement of the 2nd Bn 12th Infantry during my tour. This vast collection of facts and information has been taken further which allowed me to reconstruct the battalion deployment to Vietnam from October 1966 until its return to Fort Lewis, Washington in April, 1971 and was the core of information that propelled me to create a website dedicated to the unit's history and involvement in Vietnam. That website's URL is www.212warriors.com.

This book covers my time before I was drafted into the Army and my entire tour in Vietnam, arriving on March 13, 1968 and leaving for home March 10, 1969. I also include my final five months of service and a small review of how I got to this present day. During my tour, I was a PFC rifleman assigned to a machine gun crew as an ammo bearer, then given the job of a radio telephone operator (RTO) for 1LT Chris Brown, 3rd Platoon's leader and later 1LT R.W. *"Bud"* McDaniel. I was promoted to Specialist 4th Class or Spec Four in July and Sergeant in August when I took over 3rd squad of 3rd platoon. For a period of three weeks, I was acting platoon sergeant during that trying period in December and early January. In February, I was promoted, once again, to Staff Sergeant.

I set out to write this book in a way that would describe life as lived by many combat soldiers who fought in the area northwest of the capital city of Saigon and to give the reader a sense of what we went through day to day as well as describe some of the minutia of the how and why's of what made up the Vietnam War.

I won't go into the politics that dominated the day to day decisions that directed combat operations, large and small, and ultimately the outcome of this war. It's an offensive topic to those of us who served, to listen, read and watch, even today, fifty plus years after this war ended, pundits criticize and author judgments on what they considered the immorality of the war. There is no politics in the military. We were there to perform a duty, to serve our Country as best we could, regardless of the circumstances, of which we had no making.

FINAL THOUGHTS

In this autobiography as I portray it, there are many gaps in my memory of which I have tried to fill and the timeline may not be totally accurate, but the stories written about did take place. I did not intend this to be a full disclosure story. However, there are certain aspects of this war that need to be addressed. The American soldier did have their moments of insanity. There was the My Lai massacre which happened during my tour on March 16, 1968. You can read up on this yourself. There were other singular events of out and out murder by some troops. A Red Cross worker was killed (knifed to death) in Tay Ninh which outraged everyone. The assailant was caught and received his just due. It even came to a point where some troops were extremely disgruntled and resorted to a practice called fragging. This was a cowardly act where you tossed a grenade into someone's bunker or hut attempting to frighten or kill them. The usual victims were officers. It did not happen very often and not in our battalion, but enough to draw our attention to it.

Reports of rape and brutality were very infrequent, but this might be the case of being far removed from the mainstream life of the rear echelon troops where these activities might have occurred. Our unit was not party to any of this. On the other side of the ledger, you had NVA and VC death squads that would round up any civilian who attempted to do good and did not support their cause and they would kill them. I read that these death squads between 1957 and 1973 accounted for over 36,725 assassinations and 58,499 abductions, all done at the hands of the National Liberation Front (NLF). They singled out village leaders, medical personnel, social workers, and teachers and anyone who improved the lives of the average peasant.

This war undoubtedly has been more controversial than any other conflict we have been involved in. Political unrest, civil discourse, a popular war which quickly turned unpopular, the peace movement, a general hatred toward the military veterans of this war, the brutality of the North Vietnamese government during and after the war and the damage inflicted on the civilian population, war atrocities and the list goes on. To the veteran who fought in this war, however, the pride we have for serving our country has never been stronger and that feeling rises above all the other discourse and uneducated noise that this war produced and continues to be under academic assault. Many on the left side of the political spectrum continue to try and rewrite history and it's this same mental approach that exists today in Washington D.C. that was born out of the Vietnam War, a divided country with two different views of reality, one being born based on fact and Constitutional fundamentalism, and the other rooted in ideology for social change not truly envisioned by our founding Fathers which views our Constitution as a document that should be ever evolving to match any social changes they deem appropriate or necessary. This drive has infected our court systems to the point where liberal bias has led to judgments made to solve social issues through judicial rulings thus thwarting or overturning legislative law rather than following past precedents and the Constitution.

Scores of books have already been written justifying one way or another, the rights and wrongs of Vietnam. Try asking a Vietnam Veteran what they think about the war. The greatest percentage will tell you we fought to stop the spread of Communism and to give the people of South Vietnam the free choice of what style of government they wanted for themselves through an election process, not a military overthrow; for them to be able to enjoy the same freedoms we embrace here in the United States. Most veterans would overwhelming tell you, they would do it again...I could be wrong about this, but I truly believe if you toss out the disgruntled, which there always seem to be some, that I may be accurate in my belief that we served for a just cause. I would hate to go to my grave thinking anything else, for there was too much blood shed for it to be just a folly.

A little over two years before this book went to press, there was a lot of hoopla about a new ten part film produced and directed by Ken Burns and Lynn Novick called *The Vietnam War* (2017). It was supposed to clarify all the wrong misconceptions of the war. Instead it was another left leaning attempt to vanquish the American soldier and to further punish him for all the misdeeds America has been falsely accused of acting upon throughout history. This film was so over the top, loaded to make out North Vietnam as being the victim and the U.S. as the villain. It has been roundly rejected by the majority of Vietnam veterans with the exception of those *"sour grape"* individuals who were interviewed in the film who seemed to have a personal agenda to discredit the U.S military, themselves and history.

If you are not familiar with military terminology or need to refresh your memory, please use the Glossary in the back of the book to aid in your reading and understanding of the utilization of acronyms.

HISTORY OF THE UNIT

The Vietnam War took place in a very different time in American history. Today many liberal groups are attacking our history and even attempting to rewrite some of it because of the social injustices our country's history has written for itself, mainly at the root of this, is the subject of slavery and the long, long path we have walked as a nation, to put that behind us and to balance the scale of equal rights for all ethnic and social groups. Adding to this has been the idea of changing our language as well under the umbrella of political correctness or PC for short. This has driven us to use language which is gender neutral, and to eradicate any use of a word that might conjure up some remote thought that might cause someone to feel insulted or offended through word association. This has also been backed up by using bigotry and calling anyone a racist if one's point of view collides with an opposing point of view, the idea is to overpower and silence the other person's position as if their opinion could never be right or for that matter, shouldn't even be debated. Much of this movement, started in the early 60's through the peace movement and the usage of drugs. It has only gotten worse. Returning veterans from Vietnam found themselves the early receivers of this banter, some of it was truly ugly, being called insulting names and for some, even being spit upon. It was all the more reason why we put our uniforms in a closet and remained silent about the war for years afterward.

The unit that I was assigned to upon arriving in Vietnam was the 2nd Battalion, 12th Infantry Regiment. At the time, their nickname was the *White Warriors*, which later felt the wrath of political correctness, or maybe a bit more accurate, concerns that there might be an element of racism in the name, thus around October, 1968 we were referred to simply as the *Warriors* afterwards. Since then there have been various versions of *Warriors, Mountain, Steel* and currently, the *Lethal Warriors* are a few examples.

What about this unit's history? Any student of military history will discover that the U.S. military is steeped in tradition and legacy, not just the Marine Corp. There have been numerous redrafting and reorganizing of the military's structure over the decades. In 1957, the Department of Defense adopted CARS, *Combat Arms Regimental System*, a process designed to preserve our military history and legacy during times of military downsizing among other identifiers and rules that were laid out. There are rules that are used to guard this heritage and to keep the most storied and decorated units either active or inactive (pending recall) to ensure their history lives on. The idea is to prevent this history and legacy from being lost and ending up as part of a forgotten past.

The 12th Infantry Regiment has been formed four times in our country's history, beginning in 1798. The three former creations of the regiment are not historically linked to the present unit's history due to the rules used by the Center for Military History and the Department of Heraldry and Lineage, two departments that are housed within the Department of Defense. The three former *Twelfth's* were all *"mustered"* out of the Army, a term which means that once this was applied to a unit, it was as if they never existed on paper. They existed and yet, not, as far as the Army was concerned on paper, due to that word mustered. It's like taking an eraser to the history books and removing any reference to that unit.

1LT Charles Abbot, Jr. Adjutant, 12th U.S. Infantry, Center for Military History wrote, "The numerical designation Twelfth has been borne by four regiments of infantry in the regular service of the United States. The first was organized July 16, 1798, under an act of the same date, and disbanded June 15, 1800. The second and third were raised temporarily during hostilities, the former in 1812, its personnel being chiefly from Virginia, the latter during the Mexican War in the 1840's. Both performed well the duty for which intended, and upon the cessation of hostilities were disbanded. It was the Twelfth that was at Fort McHenry during the War of 1812, when the Stars Spangled Banner was written by Francis Scott Key.

The present regiment was organized by direction of President Abraham Lincoln in a proclamation dated May 4, 1861. An act of Congress in July of the same year confirmed the organization. It was to consist of three battalions of eight companies each. The organization was commenced in August, Major Clitz was in charge of recruiting, headquarters at Fort Hamilton, New York Harbor. On October 20th the first battalion was organized, and the return of that month shows an aggregate of 520, the companies averaging each about 60 men.

The war history of the 1st battalion 12th Infantry, indeed of the 2nd also, is inseparable from that of 'Sykes's Regulars', for the 2nd joined the 1st in September, 1862. They remained together until so reduced in numbers that the 2nd was merged into the 1st. Where ever that splendid command was engaged, the 12th Infantry did its full share. It seemed that in each major conflict the 12th found itself anchoring the middle of the skirmish line.

The 12th Regiment's first real engagement at Gaine's Mill in 'Sykes Regulars' command resulted in a total loss of 212 men, 54 of these were killed, 102 wounded and 56 missing out of a battle force of 470 strong. "Regulars" was used to differentiate between enlisted men and the volunteer forces (state militia) of the Union Army. The list of engagements or campaigns that the 12th fought in during the war is long: Peninsula, Manassas, Antietam, Fredericksburg, Chancellorsville, Gettysburg, Wilderness, Spotsylvania, Cold Harbor, Petersburg including credit for campaigns in Virginia 1862 and 1863. On November 2, 1864, the Regiment departed Virginia and Popular Grove Church for Fort Monroe, hence to New York via Norfolk with 48 men, in order to be rebuilt. The statistics of losses during the period of 1862-64 show that of all the regular regiments, the 12th stands fourth in the total of deaths including killed, died of wounds, disease, or in prison. The number that died in prison, 77, exceeds that in any other regular regiment, and indeed is one of the largest in the entire army."

Following the end of the war, the 12th Regiment completed the creation of the 3rd battalion in 1866. During the War, the volunteer units serving far outnumbered those of the regular army. The thought of enlisting and the pay that was received was not enough of an attraction to build and equip all the allotted battalions with a full complement of men. For those officers assigned recruiting duties, the struggle continued throughout the war.

After a move in 1868 to South Carolina and further southward to Alabama and Georgia during reconstruction, the 12th Infantry, beginning in early 1869, began to move west by rail and found itself beyond the frontier of the west and scattered over three states or territories, California, Nevada, and Arizona, occupying eleven different posts. They were involved in the Indian Wars of the 1870's engaging the *Modoc* Tribes, then the *Nez Perce's* in an effort to stabilize the West. They were credited with three campaigns; *Modoc's* in California, *Bannocks* in Oregon and Idaho and when they were dispatched to the

Dakota's, the *Pine Ridge* campaign. They remained in the West having visited most of the western states during this period in pursuit of the Indians. They would continue to police the west, dealing with the *Apache*, and *Sioux* nations until 1892.

1LT Charles Abbott, Center for Military History added, "In October of 1892, a detachment from the Regiment was sent from Fort Leavenworth which took part in the dedicatory exercises of the World's Fair in Chicago. While there, a gentleman seeing the regimental colors, introduced himself to the adjutant as an ex-sergeant, who had served in the regiment during the '64 campaign, and told many interesting reminiscences. Finally he produced a small package, which he carefully undid, showing a much defaced gilt star and saying—"You may look at, but not touch that. It is one of the last two stars remaining on the battle flag. In front of Petersburg another sergeant and myself cut them off, and each took one. I think everything of it." This incident indicates a lesson of devotion to the principles of duty of which the flag is emblematic, that may well be taken to heart by all whom now, or may hereafter, serve under the colors of the 12th Infantry. They will have the proud consciousness of the fact that the regiment has done its duty in the past. It *only remains for them to sustain that reputation in the future.*"

In 1898, the Regiment was ordered and dispatched to Georgia to prepare for the War with Spain. The Regiment participated in the storming of the Spanish fortress in the *Battle of El Caney* where they captured the Spanish colors in Santiago, Cuba, earning another campaign ribbon then was sent west to deal with the Philippine Insurrection of 1899 and after three more campaigns there, returned to the U.S. in 1902 for a brief period before once again, being sent back to the Island of Luzon until 1906.

Some quiet years followed for the 12th. The U.S. Army reformed itself in the second decade of the century and the unit found itself assigned to the 8th Infantry Division. The Regiment underwent rigorous training but never got involved in World War I. After a long period of peace where the 12th Regiment was stationed around the country, including Angel Island, in San Francisco Bay, they were assigned to the 4th Infantry Division in October of 1941. The Regiment's first real involvement in the Second World War was on June 6th, D-Day, 1944 as they waded ashore at Utah Beach along with their sister battalions, the 8th and 22nd Infantry Regiments. Their first task was to relieve elements of the 101st Airborne Division who parachuted in behind enemy lines to secure the cross roads at St Lo. After this was achieved, they made a pivot northward to secure the Cherbourg Peninsula and its harbor, a required need to resupply the invasion.

After the taking of Cherbourg, the 4th Infantry Division moved south through Periers, then changed course on an easterly heading to Saint Eny, Saint Lo, then through hedgerow country on toward Saint Pois. It was around the 1st of August, the French 2nd Armored Division, trained by MG George Patton in England, landed at Normandy. They were under the command of General Phillippe Leclerc, who got his combat experience while in North Africa, commanding L Force in Libya in 1943 and covered the British Eighth Army inland flank during the advance on Tunisia. In short order, they joined LTG Omar Bradley's 1st Army and the 4th Infantry Divisions march through France. After a pause in Le Teilleul area, they moved to Arpajon, about 10 miles outside of Paris. At this point, the 12th Infantry was the lead element for the 4th Infantry Division. Supreme Allied Commander Dwight Eisenhower had promised Leclerc back in 1943 that the French would be the ones to liberate Paris and it was around this point that the French 2nd Armored was assigned to lead the approach to Paris. The French struggled to move forward primarily because they were stopping to party at every French village or farm house while the 4th ID was being strafed by the Germans. This went on with the French for two or three days. LTG Omar Bradley, commander of the U.S. First Army (and later 12th Army Group) was getting very pissed. Leclerc was promised by Eisenhower that he would be the one to free Paris, yet at this point in time, he was not helping his own cause. Hearing from Bradley that the French are holding up the parade, Eisenhower in frustration says "To hell with the French, send in the 4th". Bradley ordered his lead Task Force elements of which the 12th Infantry was part, to advance and take Paris. They were the first American unit to enter the city from the south, and when the French realized what the Americans were doing, headed toward the city at breakneck speed and entered Paris from the North in an attempt to save face. Two days later, historical photos show the 28th Infantry Division marching through the Arc de Triomphe and down the Avenue des Champs-Elysees as if they were the liberators of Paris, but it was really the 4th ID. The reason the 28th ID was doing the marching was because their uniforms were cleaner and that was the only reason, while the 4th ID, who had already exited the city was pursuing the Germans as they fled the city.

As the 28th ID marched in Paris, the 4th ID moved northeast toward Chauny to Margut, then crossed into Belgium and took up a position at St. Vith, north of Bastogne. They later were sent to the Hurtgen Forest, a meat grinder that tore up the 28th ID and had no mercy on the 4th ID either. After this beating, which lasted from the 9th of November until December 3rd, when the 4th ID was relieved after suffering over 5,000 casualties, were sent south to regroup to an area northeast of Luxembourg City, Luxembourg. The 12th was assigned an area of the front line just outside of Echternach (Ester-knock) where they were to anchor the southern flank of the American lines. This was supposed to be a quiet zone, to rest and get healed. Two weeks later, the regiment was fighting for its life. They were to be awarded the Presidential Unit Citation for their involvement in the Battle of the Bulge defending Luxembourg from the German Army. The 12th Infantry Regiment took the brunt of the 212th German Division's assault when they crossed the Sauer River and were instrumental in protecting the American's right flank, which allowed General Patton's 3rd Army to counterattack the Germans. After this battle, they crossed the Sauer River into Germany and pushed southeast bypassing Heidelberg on a course that would carry them across the Danube River and place them below Munich and just above the Bavarian Alps where they are relieved by the 101st Airborne in April 1945, then soon after, were sent home to prepare for the invasion of Japan. The dropping of the two atomic bombs on Japan made that unnecessary. They were credited with five campaigns during the war.

After Belgium gave the 12th Infantry the Belgian Fourragere, it was written somewhere in the language of the citation that the regiment must have been led by love of country to carry out its mission in defending Belgium and Luxembourg. It is from this citation that the Regiment adopted its official motto, Ducti Amore Patriae, *"Having Been Led by Love of Country."*

The Regiment was assigned to the 4th Infantry Division in 1963 and in October 1966, arriving by troop ship aboard the USS Walker, found itself wading ashore at Vung Tau, Vietnam. The Regiment was the only unit to have five battalions fighting in Vietnam and earned 11 campaign ribbons plus three Presidential Unit Citations before they returned home in April of 1971. After the Vietnam War ended, the 2/12th found itself being inactivated until 1989 when it was reactivated at Fort Carson, Colorado with the 4th ID. In 1995, the unit was inactivated again and relieved from assignment to the 4th ID.

After another long layoff, the 12th Infantry Regiment was once again called into action in 2005 and assigned to the 2nd ID at Fort

Carson and in 2006, was sent overseas fighting terrorism while serving in both Iraq and Afghanistan theaters, being deployed numerous times to each country which continued up to November 2018 in support of Operation Iraqi Freedom and Operation Enduring Freedom (Afghanistan). In 2008, the unit was reassigned from the 2nd back to the 4th ID. The *Global War on Terror* campaign continues and the 12th Infantry Regiment remains a vital and active element of the 4th ID as the *go to* Division of the U.S. Army while stationed at Fort Carson, Colorado Springs, Colorado. As of this writing, the 2nd Bn 12th Inf has become the most deployed unit in the Army, conducting its eighth overseas mission between 2006-2018.

The 12th Infantry *Warrior* Regiment is a highly decorated unit. During its long history, it has fought in seven wars and has been credited with four Presidential Unit Citations, six Valorous Unit Citations, eight Meritorious Unit Citations, Army Superior Unit Award, the Belgian Fourragere 1940 and three Vietnamese Cross of Gallantry with Palm Unit Citations, Republic of Vietnam Civil Action Honor Medal, First Class and 39 campaign ribbons. The Regiment currently has seven Medal of Honors awardees. Among those seven mentioned is SP4 Donald W. Evans, age 23, a medic attached to A Co 2/12th, 4th ID at the time, who received this honor posthumously, after being repeatedly wounded by small arms fire while coming to the aid of numerous wounded during a battle on January 24, 1967. During this engagement, Donald refused to seek medical treatment for himself and after treating another comrade, was mortally wounded.

In January 1966, the 4th ID was put on alert that they would be heading to Vietnam. The 4th ID was located at Fort Lewis, Washington. The 2/12th was depleted at that time, so three battalions were quickly formed that began immediate training. All the troops were new, almost exclusively draftees, with no combat experience in the ranks, including the officers and NCO's that formed the battalion.

This statement is based on interviewing many of the officers and men who served with the unit that was initially deployed to Vietnam. Only the senior officers had combat experience from WWII or Korea. The men trained as a unit, from basic to advanced individual training (AIT). The 1st and 3rd battalions of the 4th ID were the first to be sent over by ship and were stationed near Pleiku in the central highlands. The 1/12th attached to the 2nd brigade arrived on August 6th and the 3/12th attached to the 1st brigade arrived October 4th. They joined the 3rd brigade, 25th ID which had arrived there in December, 1965. The rest of the 25th ID, the 1st and 2nd brigades had arrived earlier in January and April, 1966 and were deployed in the south of Vietnam to III Corp area located around Cu Chi.

The 4th ID's last brigade, the 3rd, left for Vietnam on September 22nd, arriving at Vung Tau, Vietnam, by ship via the U.S.S. Walker, on October 10, 1966. The 25th ID needing more troops to secure their northern area of operation around Dau Tieng benefited from the arrival of the 4th's 3rd brigade, which included the 2/12th and found themselves being attached to and under the operational control of the 25th ID.

The brigade moved from the beaches of Vung Tau up the highway to Bearcat, later named Camp Martin Cox after SFC Martin Cox, a combat engineer who was killed on April 7, 1966. That area was part of the 1st Division's AO, area of operation. The 1st ID had been working north and west of Saigon and had arrived there in July of 1965. Their task was to prepare and clear out areas for other units to arrive and get set up and settled in. Once the 3rd Brigade, 4th ID arrived in country, it was the 1st ID that provided some in country training to the unit. While the 2/12th and the rest of the brigade received some indoctrination training about their adversaries, the 1st Brigade of the 1st ID was sent to Dau Tieng to clear out the local VC so when the 3rd Bde, 4th ID arrived they could build the basecamp undisturbed.

After a month's stop at Camp Cox, also called Bearcat southeast of Saigon, the 3rd Brigade was sent to Dau Tieng by truck convoy through the streets of Saigon, to the former headquarters of the Michelin Rubber Co and where they operated an extensive rubber tree plantation. They were assigned to carry out Operation Nisqually, and build Camp Rainier next to the village of Dau Tieng, a location in need of a permanent combat presence, where they could operate out of and conduct combat operations in the region. This was done between November 1966 and January 1967.

So, the stage was set for two brigades to be operational controlled (OPCON'd) by another division for almost a year before it was decided to address this situation. On August 1, 1967, both brigades were reflagged, meaning the 3rd brigade of the 4th ID and 3rd brigade in the 25th ID changed their division patches and on paper, exchanged places with each other. This is how the 2/12th was assigned to two different divisions in the Vietnam War. In 1967, two more battalions were trained in Fort Lewis and deployed to Vietnam, the 4th and 5th battalions which were assigned to the 199th Light Infantry Brigade, making the 12th Infantry Regiment the only unit to have five battalions in the war.

One of the largest battles of the war took place at LZ Gold or FSB Gold, also known as the Battle of Suoi Tre. This battle involved the following ground units that all played a major part-part listed by unit, battalion C.O. and their call signs:

3rd Bn 22nd Inf—LTC John A Bender (Falcon 6)
2nd Bn 22nd Inf (M)—LTC Ralph Julian (Fullback 6)
2nd Bn 34th Armor—LTC Raymond L Stailey (Dreadnaught 6)
2nd Bn 77th Artillery—LTC Jack Vessey (Focus 6), Later became 4 star general & Chief of Staff for U.S. Army
2nd Bn 12th Inf—LTC Joe F Elliott (Flame 6)

In the morning of March 20, 1967, the 2/12th moved northwest of FSB Gold on a Search & Destroy operation into AO Silver with each company running a separate azimuth. The 2/22nd was operating southeast of LZ Gold along with the 2/34th tankers about a click away and south of the Suoi Mamat River.

C Co veteran SGT Les Cooper recalls, "Prior to us going in to the LZ [on the 19th] we had lost a couple of choppers. There were command detonated mines (or least that was what we were told). We finally arrived at the LZ; dug in; spent the night and then moved out the next day."

C Co commander CPT John Napper, who just days before had replaced CPT Cris Stone, who led the company at Fort Lewis, Washington in training, and deployed with the unit recalls, "After spending the night at LZ Gold, the battalion (BN) headed out early in the morning. Yesterday, brigade commander COL Marshall Garth had spotted a group of 40 VC on the move from his command chopper and today wants the BN to pursue this activity. As we moved out toward that area, we discovered a fresh basecamp with cooking fires still going, hot rice ready to eat, and a water buffalo that was left behind. LTC Joe F Elliott, the BN commander ordered us to take the buffalo with us. He had some idea of exchanging it with some village to gain favor. A G.I. from Wyoming, a cowboy at that, was assigned to drag the buffalo along with us. After failing to find the enemy, at the end of the day we set up a night laager with the rest of the BN out near a wooded area NW of FSB Gold about 2.5 clicks (a click is 1000 meters) away. The BN built their night defensive positions. There was no concertina wire. All was quiet that night, including the water buffalo which had survived the journey."

SGT Earl Noble Jr, B Co recalls, "We had been out on a sweep and stopped to rest. Off to our right, there was a TAC air strike taking place. They were hitting a basecamp in the distance. As I was watching, I heard a swoosh, swoosh, swoosh coming closer and louder. A G.I. sitting about 10 feet from me all of a sudden gets hit in the head with a chunk of shrapnel. He had to be Medevac'd out."

The next day, March 21st at dawn, LTC Elliott held his morning briefing on the operational plans for the day. The company commanders headed back to their respective assigned areas of the laager site to brief their platoon leaders. At approximately 0631 hrs, the VC fired a number of rockets or some type of HE rounds at the laager site about the same time as they began their attack on FSB Gold in the distance. There were wounded and SP4 Larry Barton, from A Co, was killed from the volley. There seemed to be some dispute by the guys in various companies whether the rounds that hit the laager site were fired by VC or whether it was done by our own artillery. You could clearly hear the firing in the distance from the southwest which was the location of the nearest supporting FSB and moments later, the rounds hit.

Maybe they had their fire mission coordinates wrong and they should have been firing support toward FSB Gold. By 0655 hrs, the day's orders were changed, the laager site was broken down, and the Bde Commander informed LTC Elliott to make haste back to FSB Gold as they were under a heavy attack. Charlie Co which was occupying the SE section of the laager took the lead heading out in single file with flankers out. Just prior to this, according to CPT Napper, two machine gunners were assigned the duty of taking care of the buffalo. They say the hell with this, and as the company was leaving the perimeter, the buffalo becomes a casualty of the war. Alpha Co followed after Charlie Co at 0735 hrs with HHC and LTC Elliott as part of the formation. Bravo Co remained behind along with HHC Recon to tend to the dead and wounded, then they moved out toward FSB GOLD about 0840 hrs. On the way toward FSB Gold, C and A Co's were mortared twice, but the rounds fell short.

Around 0900 hrs, A Co arrives at the southwestern edge of the clearing 600 meters from the edge of the FSB and CPT Allyn Palmer directs his men to assault the VC in a cross fire. They moved forward and provided relief to the beleaguered B Co 3/22nd with everyone firing away at the enemy. To their left and SW was Fullback Charlie 6, call sign for C Co 2/22nd (M) which had arrived from its night laager site to the South. As A Co 2/12th moved forward, the 2-34th burst out of the tree line to the southeast with cannon and machine guns blazing, about the same time as the 2/22nd passes through the ranks of the 2/12th. The enemy was hit hard from two sides. CPT Palmer and his company teamed up with B Co 3/22nd then they counterattacked the enemy and retook their forward positions around the southern perimeter.

C Co came into the clearing about 30 minutes ahead of A Co, from the North side of the FSB and swung around the perimeter toward the east and south, pinching down on the withdrawing VC being hit from the other side of the perimeter by A Co 2/12th and 2-34th Armor. The VC were now in full retreat, heading east away from the laager site. CPT Napper met up with LTC Elliott and in the process asked Napper where he's been. Napper replied, "I've been here for 30 minutes." As C Co approached the firefight, CPT Napper noted, "It was a hellacious firefight listening to it from a distance. It was one of the most confusing 3-4 hours that morning. As we broke into the clearing, the 2-34th was firing at everything. I needed to let them know there were friendly's out there. As I stood up, I saw the barrel of a tank swing in my direction. As I hit the ground I stuck the barrel of my rifle in the ground and the canister round took the forward stock of the rifle clean off. I was left without a weapon."

Within 45 minutes of the arrival of the infantry and mech units, the attack had been crushed and mop up began. There were dead everywhere. The mech units pursued the enemy into the woods. Bravo Co arrived soon after, but the shooting was over.

C Co veteran SGT Les Cooper wrote, "We were not mortared. They had spread each company out—we were the bait...we did the drive as did 'A' company. I can remember a radio call to the 'LZ' notifying them we were coming in. I was in the heavy weapons squad for 1st platoon; we were so short of people I had volunteered to carry the M60. I remember entering the LZ and tossing a hand grenade at the quad fifty—it had been over-run. Other than that, I think I fired four or five hundred rounds, not sure I hit anything but it was supportive fire. You might check the number of wounded, number was increased by 'friendly fire'—no fire is friendly. I know the ones we had were."

In the aftermath of the battle, the BN Chaplin was upset about the treatment of the VC soldiers, both alive and dead, according to CPT John Napper."He didn't like the idea that the mech and tanker boys had run over bodies in the heat of the battle, some alive at the time. But what would you do when you're on a rescue mission trying to save U.S. lives? He kind of softened his statements after reflecting on the issue."

A Co 2/12th veteran Bill Comeau wrote of the reception that the Warriors received when they entered FSB Gold. "Artillerymen...ran out of their foxholes and threw their arms around us with tears in their eyes."

"Presidential Unit Citation is awarded by the direction of the President of the United States to THE 3D BRIGADE, 4TH INFANTRY DIVISION and Assigned and Attached Units FOR EXTRAORDINARY HEROISM:

The 3rd Brigade, 4th Infantry Division and the Attached and

Assigned Units distinguished themselves by extraordinary heroism while engaged in military operations on 21 March 1967 in the vicinity of Suoi Tre, Republic of Vietnam. The members of this Brigade and the foregoing units demonstrated indomitable courage and professional skill while engaging an estimated force of 2,500 Viet Cong. During the early morning hours of 21 March 1967, an estimated force of 2,500 Viet Cong launched a massive and determined ground attack against elements of the 3rd Battalion, 22nd Infantry and 2nd Battalion, 77th Artillery located at Fire Support Base Gold near Suoi Tre, Republic of Vietnam. Opening the engagement with an intense mortar attack, the enemy force, later identified as the 272nd Main Force Regiment reinforced by two additional infantry battalions, struck the perimeter in three separate locations.

Due to the ferocity of the assault and the overwhelming number of enemy troops, untenable positions in the north and southeast were overrun within the first 30 minutes of the battle, despite determined resistance by friendly forces. As the enemy penetrated the perimeter, the American troops set up an internal perimeter and continued to direct withering fire on the enemy. When the Viet Cong directed anti-tank fire upon the artillery positions, heroic gun crews cannibalized parts from damaged guns, and, at several points, fired directly into the advancing enemy, including the firing of beehive ammunition through gaps in the perimeter.

While the battle continued to rage and grow in intensity, the Brigade Commander was directing the 2nd Battalion, 12th Infantry, the 2nd Battalion, 22nd Infantry (Mechanized) and the 2nd Battalion, 34th Armor, to the besieged fire support base. At the same time, the support and service elements of the brigade began a furious aerial resupply of ammunition and medical supplies from the Brigade rear basecamp at Dau Tieng.

As the 2nd Battalion, 12th Infantry began its overland move to the fire support base approximately 2,500 meters distant, a heavy concentration of enemy mortar fire was directed upon their position, killing one man and wounding 20 others. Undaunted, the battalion moved nearly 2,500 meters in less than two hours despite constant blocking and harassment efforts by the enemy. Concurrently with the movement of the 2nd Battalion, 12th Infantry, mechanized and armor elements began moving across the Suoi Samat River at a ford which had only recently been located and which previously had been thought impassable.

Driving towards the fire support base, the mechanized unit, followed by the armor battalion, drove into the western and southern sector of the engaged perimeter passing through engaged elements of the 2nd Battalion, 12th Infantry. Striking the Viet Cong on the flank, the 2nd Battalion, 22nd Infantry smashed through the enemy with such intensity and ferocity that the enemy attack faltered and broke. As the fleeing and now shattered enemy force retreated to the northeast, the 2nd Battalion, 34th Armor swept the position destroying large numbers of Viet Cong who were now in full retreat.

Throughout the battle, fighters of the United States Air Force, directed by the Brigade's Forward Air Controllers provided close support to the fire support base and hammered enemy concentrations outside the perimeter. As the FAC aircraft dived through heavy anti-aircraft fire to mark enemy positions, the plane was hit by ground fire and crashed, killing both occupants.

After securing the fire support base, a sweep of the area was conducted, revealing a total of 647 Viet Cong bodies and 10 enemy captured. It is estimated that an additional 200 enemy were killed as a result of the aerial and artillery bombardment. Friendly casualties were extremely light, resulting in only 33 killed and 187 wounded of whom approximately 90 were returned to duty.

Through their fortitude and determination, the personnel of the

3d Brigade, 4th Infantry Division and attached units were able in great measure to cripple a large Viet Cong force. Their devotion to duty and extraordinary heroism reflect distinct credit upon themselves and the Armed Forces of the United States."

Combat effectiveness was an early challenge for the division because of being deployed as a unit. Every soldier was due to come home about the same time from their one year tour of duty and this needed to be addressed. So, the 4th ID began to swap out soldiers to other division units and bring in both replacements and soldiers in order to diminish the combat experience losses due to rotations home. It didn't make the average Joe happy, but it worked.

This constant churn that occurred put tremendous pressure on the officers and NCO's who would gain rank and leadership only to depart and then the cycle would start over. There was a constant change to personnel and that required the replacements to gain experience fast or die trying. In the year that I was an infantry rifleman, according to the records, our average company strength was about 100 men. During the period of March 1968 to March 1969, almost 400 men came through the company. This was due to death, wounded, sickness, transfers to other jobs, and rotations home. No wonder many of us could not remember the names of those we served with.

One last bit of historical fact before I begin my story. When the 2/12th was stationed in Alaska during the early 1960's, they were given a totem pole by the Native Americans. This pole followed the unit around from post to post.

Depicting major conflicts in the history of the United States in which the 2/12th has taken part, the *Warrior* totem pole was situated in front of battalion headquarters in Dau Tieng, next to the airstrip. Unveiled in a brief ceremony in May of 1967, the newest addition had been placed into position atop the 40-foot structure. It depicts a North Vietnamese Regular, symbolizing the unit's participation in the Vietnam Conflict.

Long before the unit's arrival in Vietnam, the totem pole sat at Fort Lewis, Washington serving to remind all incoming recruits of the outstanding performances they were called upon to continue as an integral part of the 2/12th's fighting forces. Each figure carved in the pole represents the unit's participation in a major conflict, beginning with the War of 1812, proceeding through the Mexican War, Civil War, Indian Campaigns, World War II, and now the conflict in Vietnam.

When the unit deployed to Southeast Asia, the pole was left behind at Fort Lewis, Washington. Normally, it is the duty of the senior battalion to pack and store all trophies, photo albums, etc., safely away. In this case, that was the 1/12th. They didn't do it so the 2/12th was asked by Division to perform the task. Responding to a challenge from his fellow officers and answering to a bet that he couldn't get the pole to Vietnam, BN CMDR COL Marvin D. Fuller had the pole cut into segments, carefully packed and crated and shipped to Vietnam. It was not until the unit reached Dau Tieng and Camp Rainier that the pole was reconstructed. They added an NVA soldier to the top of the pole to represent the current conflict.

A large combat infantryman's badge rests near the top of the structure, symbolic of the American infantryman in all wars. The totem pole stood as it did back in the States as a tribute to those who have served so faithfully in the 2/12th Inf since its beginning. So, what happened to the Totem Pole when the unit left Dau Tieng, and later in 1970 when it became a *brigade separate* unit operating under the command of II Corp (pronounced 2 core) east of Saigon at Camp Husky? Brigade separate meant they were not assigned to a division at that time and operated as an independent brigade. Apparently the totem pole was moved down to Cu Chi after the 2/12th was moved over to the 2nd Brigade in late 1968. Someone lost track of the significant symbol of the totem pole because it was never returned to the

U.S. Whether it was taken by local villagers, VC, or destroyed by the communist foot soldiers we may never know. So much for my unit's history...at least, this particular contribution.

The totem pole story was provided by COL Marvin Fuller and Bill Comeau, RTO for CPT Allyn Palmer, A Co, 1966-67.

We begin our story

Young Men in Harm's Way

1. OPERATION WILDERNESS

Sleeping on the ground was not new to me, but where I was, it's the new norm. Finding any comfort in rolling around on rocks and over indentations that poked me in the back and sides, especially when they were rock hard, had left me with a listless night's sleep. I stood up and stretched my arms up over my head and looked around. I was among strangers who were all dressed in green. I was trying to sort out what was expected of me. There are rules that need to be learned quickly. Slowly activity and the movement of people began to take on purpose. Men were shouting to those around them. It was early as dawn breaks and the first rays of light began to filter through the trees to our north. The air was dry and there was still a measure of coldness to it. There would be no hot breakfast this morning, or a chance for a cup of coffee. The sounds that woke me from my sleep were the volleys of rifle fire from first platoon as they sprung their night ambush on a group of Viet Cong.

I asked PFC David Schultz, "What day is it?" and he replied, "April 4th, remember?"

I paused for a second, "Oh yeah," as I tried somehow to clear the fog from my brain.

SGT John Spoores came over to check on us, giving pause to our conversation as brief as it was. "First platoon went out on ambush last night," he started out with, "and they reported seeing campfires and hearing conversations about 100 meters from their position. They said it looked like there were a lot of gooks up there. They were only a squad, six or seven guys so they thought it wise to keep a low profile. As soon as the sun came up they decided they needed to return to our laager site before they were discovered," Spoores said. He then continued on, "On the way back they spotted a couple of VC that got in behind them and quickly set up an ambush and then sprung it on them. They hustled back here right after that happened. Word has it that we will be breaking out of our laager site at 0800 hrs. Get your shit together and be ready to move out."

SGT Spoores, like the rest of us, was slightly above being a baby-faced kid and had been in Vietnam since September 1967. He had already seen plenty of action. As my tour of duty began in the Republic of Vietnam, there were two things most of us will remember about the men around us and that is their last name and what state they were from. John was a Hoosier and hailed from Indiana. I would guess he's 22-23.

About 15 minutes later, Sarge came back and told the squad to huddle up. He said we will be moving out and that we will string out and get online so that we are spread out between the APC's, the M113 armored personnel carriers. *Command*, in this case was our battalion commander, LTC Charles Bauer, wants us spread out so that we can cover the entire area of jungle to our front and not allow any enemy to slip away or through us. For the past several days, we had been working under the control of the 199th LIB (Light Infantry Brigade). Attached to the 199th was the 3/17th, 3rd Squadron, 17th Infantry (M), M for Mechanized, an armored unit of M113 APC's (armored personnel carriers) or *tracks* as they were nicknamed.

"Saddle up! Get your gear on, we're moving out," Spoores said.

I'd been assigned to one of two M60 machine gun (MG) crews in the heavy weapons squad, also referred to as fourth squad. I grabbed my backpack frame which had 200 rounds of 7.62mm MG belted ammo, four canteens of water attached to the frame and swung it over my shoulders and adjusted the straps. I grabbed my pistol belt with its two attached magazine pouches and buckled it around my waist. I reached down and picked up my brand new M16A1 rifle. I was ready to go.

As I looked up, Spoores yelled out, "Lock and load."

I hit the release button and quickly caught the 20-round magazine from my M16, then cradled it in the palm of my hand, tapped it several times on my thigh to ensure the rounds were moving freely, then slammed the magazine back into my rifle and listened for the click. Then I grabbed the charging lever and pulled back on it until it stopped, then I released it in one quick motion and let it fly as it slammed a cartridge into the rifle chamber. I double checked to make certain the safety was set to the on position. The sun by now was still low in the sky. Last night, normal procedures had the F.O. (Forward Observer) dial in artillery supporting fire prior to the sun going down. The F.O. was usually an officer, a lieutenant from one of the attached artillery batteries from our Brigade. Every company had an F.O. traveling with them on all operations. In this case, we were being supported from a Fire Support Base or FSB for short. Nearby was FSB Hampton, right on the outskirts of a small village called Go Dau Ha. The village lies next to the Song Vam Co Dong, Vietnamese for Vam Co Dong River, one of the larger rivers in this section of III Corp. Those who have not served in Vietnam during the war will not have a clue where any of the landmarks mentioned are and maybe that just isn't important.

We were located about 45 miles northwest of the capital city of Saigon and 10 miles due south of Tay Ninh. At the FSB was a battery of 105mm howitzers from the 2nd BN 77th Field Artillery (FA). The F.O. called for a fire mission to the fire direction control (FDC) officer who relayed the map coordinates to the gun crews. Our F.O. asked for a WP (white phosphorus) marker round set with a timing charge for map grid coordinates XT214445, the location of our NL (night laager) site. Laager is a term to describe a defensive position that is set up to protect ourselves against the enemy at night. As the F.O. called in each marker round, high overhead the explosion of each WP round nicknamed *willy peter*, can be seen as a pure white cotton cloud forming a ring in the sky over our position. By doing so, if the unit came under attack at night, the F.O. could call in pinpoint accurate artillery fire on the enemy assault without having to adjust the artillery fire, thus saving precious time and lives.

Our NL was set up in the middle of a large rice paddy, hard as a rock from the heat of the dry season which runs from August to April or so. There were three companies of infantry from the 2nd Battalion, 12th infantry (hereafter 2/12th), 25th ID. B Co was under the command of CPT Harry Holter, C Co commanded by 1LT Jay Hickey and D Co led by CPT Ed Bethea. There was also a mech infantry unit, D Troop 3rd Squadron 17th Calvary Regiment (D 3/17th Cav). All told, we were about 320 men strong. Word filtered down via radio for an officer's call, and in turn, the C.O.'s told the platoon leaders who told their NCO's what the strategy would be this morning, knowing the enemy was close at hand. The orders we got will put young men in harm's way today, an online assault. Charlie Company, my company was told to string out single file to the left flank, and Bravo Company was on our right flank. To our rear, in reserve was Delta Company. We walked out and got into position, standing in a long line out in the rice paddy with the green foliage of the jungle to our front. Across this long line of men interlaced amongst them are eight APC's with their .50 cal machine guns. Once everyone was in position, over the radio the announcement to *"Move out"* was heard and then we saw the hand signals from our PL (platoon leader), 1LT Hugh Vandervoort.

About 0900 hrs as we were moving cautiously into a heavily wooded area, all eyes were sweeping the landscape for any kind of movement or any shapes or lines that didn't resemble natural habitat. The canopy of the forest created mottled patterns of shade and sunlight on the bushes and tall grasses that existed far below the tree tops. This odd existence of lights and darks concealed the danger that lurked ahead. My eyes scanned back and forth, jumping to imaginary figures that turned out to be nothing but different shapes of vegetation resembling a human torso. There was a sense of tension that could be felt all around. A few minutes later, one of the men called out that he saw *bodies*. Up ahead two VC had been hit by harassing artillery fire from the previous night, another trick we used to discourage the enemy from congregating in an area to amass a ground attack. These two were dressed in green fatigues, no hats and Ho Chi Minh sandals, strips of auto tires bound to the feet with string, rope or vines tied to hold them onto the feet.

Specialist Fourth Class (SP4) Dixon was searching the bodies for papers or anything he could find. He stripped one body of its wallet, watch, and checked some photos that we found, and then tossed them to the ground. The other body was searched as Platoon Sergeant (PSG) Tommy Knapp's radio telephone operator (RTO) radioed over to the Lieutenant what we had found, then relayed this information back to Charlie 6, the call sign for 1LT Jay Hickey, our C.O. Jay was from Georgia and had just taken over the command of the company a few weeks earlier. He had his shit together and was well respected by the men. His RTO was PFC Hiram Marziano, a tall lanky kid with jet black hair, a southerner from North Carolina, following his every movement, never more than a step or two away from the C.O. Most

C.O.'s spent three to six months in the field before being assigned to HQ or HHC (Headquarters, Headquarters Company), the officers and EM's (enlisted men) that make up the support staff for the battalion commander.

I'm a FNG (f**** new guy) and as such, my first name will soon be stupid, but right now, I saw what was going on around me and I was thinking to myself, "Hey, I want in on this action." This was not a good thought as it turned out. I ended up tossing caution to the wind, but I didn't know that yet. We got orders to continue with our sweep of the jungle. As we moved out, the APC nicknamed *Jumpin Jack Flash*, pulled up along the side of us then slowly moved out in front of us about 25 meters before stopping. The driver idled his diesel engine as the track commander scanned the bushes in front of them as they waited for the infantry and my platoon to catch up to their location. As we began to move in front of them, we started the process all over, leap frogging our way forward. Close to 0930 hrs word came down to take 10. It was smoke break time so I found a clear spot on the jungle floor and got off my feet.

As I lit up a cigarette, my mind drifted back to when I was a kid, around 10, I would guess. I grew up in Santa Rosa, California out in Rincon Valley, a small farming area just four miles east of the downtown. There were lots of open fields, hay and weeds, prune orchards, and sheep pastures to wander around in, whether the owners of the land liked it or not. My buddy Billy Hougen and I would venture away from home for hours at a time exploring this large world of ours, unbeknown to our parents. Today, we would probably be grounded for a month for this activity, but back then, it was a much simpler life. For as long as I can remember I would wear cowboy outfits, with the hat, western shirt, and two pistols on my gun belt. Playing Cowboys and Indians was one of my favorite pastimes. I watched the Lone Ranger, Roy Rogers, Lash LaRue, the Cisco Kid, and Range Rider, among others on TV. To top off this love for shooting my fellow man, my biggest hero was John Wayne starring in *The Sands of Iwo Jima*. Yep, playing war and cowboys and Indians, and now I find myself in a foreign country living out this fantasy, the only difference, as I have this conversation with myself, is that I could get killed for real here. "On your feet!" I hear a friendly voice belt out, that jolts me back to reality.

The APC engine revs up behind me and a few seconds later the APC slowly maneuvered past my location and I got a snoot full of black diesel exhaust. I glanced down at my watch, its 0950 hrs, then I made eye contact with Schultz and he smiled back; two dumb FNG's, about to make a rookie mistake. Later on in my tour, I would remember this moment and use it as a training lesson to others who would be under my command.

During the morning's activities, I don't recall separating myself from our gunner, but we must have. As I think about this years later, that was a cardinal sin. Our primary responsibility was to provide support for our machine gunner. Anyway, by us, I mean David Schultz and myself. I don't remember how long we had been in this formation and how far we had advanced from our laager site. I'm guessing it could have been 1,000 meters or so. Not a great distance at any rate. Suddenly, one of the track commanders stopped and signaled the rest of the tracks to stop. We had eight APC's on line at this point.

David and I made eye contact with the guy on the track and he signaled us to come over. I noticed the stripes on his fatigues as he looked at me, and then he indicated with a hand motion that he had picked up on something.

"I think I see movement to my front beyond that row of trees. I'm not moving my track any further," the track commander said, displaying sergeant stripes on his sleeve.

"What do you want from us?" I asked, as I glanced over toward Schultz.

The geography we had been going through consisted of areas of open grass and surrounded by trees and underbrush. At that point we were kind of staged behind a hedgerow, which was 10 meters in front of the tracks. Hedgerows can be trees, brush and bushes which create a makeshift natural fence line. These were often used and found along property lines. There was some kind of a road or oxcart trail beyond the hedgerow that crossed in front of our unit's formation. Beyond the road was a heavy tree line that stretched the width and beyond of our troops and tracks.

The Sergeant continued, "Someone needs to go up there and check the situation out." I was thinking to myself, "Hey dude, you got the armor and a .50 cal. Why aren't you cozying your APC up there yourself?" but those words never got to my mouth.

It was at that moment in time where that "throw caution to the wind" idea overcame my training and whatever limited experience I had managed to acquire in the eight previous days I had been assigned to this unit. I felt somewhat giddy about the situation. I was still thinking about finding those bodies earlier, and wanted to do a little of my own discovery. Schultz and I were clearly not up to the task and were out of position.

"OK, we'll move up and see what's going on," I said, without any regard to the perilous situation David and I were about to engage in.

We left our gunner and wandered up past the APC's and to their left. We stepped through a break in the hedgerow and walked up to the edge of the trail, then onto it. To my right was the oxcart trail and I could see down it for several hundred meters. Directly to my front, the trail made a 90 degree turn and ran away from my position.

Schultz was standing to my right about 10 feet away and we were both surveying what was to our front. I glanced back to see what the rest of the company was doing. I peered through the branches of the bushes that were between us and the tracks. They seemed to be holding their positions, APC's included. We must have been ahead of them by 40-50 feet. I nervously peeked at my watch, it was 0955 hrs. Suddenly, all hell broke loose. There were AK-47's popping off and the angry buzz of bullets were whizzing by me. I looked to the right to pick up the location of the firing as I dived for the ground. As I was falling to the turf, I saw Schultz grabbing his midsection and then he went down hard and never moved. He had taken the initial burst from the small arms fire. We had walked up on an NVA (North Vietnamese Army) basecamp.

The sounds and bursts of weapons fire was everywhere. I could hear the rocket propelled grenades (RPG's) being fired, followed by explosions. The level of sound in a battle was fierce. You had to yell to be heard. The APC's began to exchange fire as they opened up with their .50 cal.'s, *TAT-TAT-TAT-TAT…TAT-TAT-TAT,* the slow but powerful sound of each round being fired as they started to back away from the intense fire coming from the hedgerow across the road and to my front. The engines of each track were being gunned to full throttle as they slammed the transmission into reverse. I looked again to my right and could see a number of soldiers in bunkers firing to their front. I flicked the safety off my M16 and unloaded a clip in their direction as I took aim, setting the site of my rifle barrel on several pith helmets that were bobbing around in the nearest bunker. I needed to get over to Schultz to see what I could do for him. He was still not moving nor talking. We were both carrying 200 rounds of 7.62 for the M60 plus our own equipment. Schulz was no small kid. I figured he had to weigh around 190 lb and the equipment on him added another 30-40 lbs.

Wasting no more time, I crawled over to check David for wounds. "Schultz, can you hear me? Can you tell me where you were hit?"

He didn't respond and was motionless. I continued to run my hands over his body searching for blood. It was difficult considering I was trying to do this lying flat on the road surface as I made my body as small a target as I could. I was feeling his body for life signs and blood but I couldn't find any and, try as I might, I couldn't feel a pulse either. I was receiving more ground fire from the tree line and the bunkers. Bullets were dancing on the ground toward me. I had no fire support and I was lying with David on the oxcart trail, exposed to the enemy. My unit had pulled back. My gut reaction was that David is gone. I knew he took it across the midsection from his right, probably through the liver and stomach, but I wanted to be sure. I tried to roll him over but had trouble doing that with his pack still on. There was another NVA soldier firing at me so I grabbed another clip and fired back, trying to suppress the small arms fire from the closest bunker. Schultz was still unresponsive. I reached around him one more time and as I was doing this, I was hit in the upper arm by an AK-47. That sealed the deal for me. It was getting too heated out here in the open. I couldn't drag Schultz to safety even if I wanted to at this point, not without the suppressing fire I needed in order to stand up and gain leverage. I stood no chance against those automatic weapons out in the open.

I had to get help in the hopes that Schultz was only unconscious and not gone. I didn't want to leave him, but at that point me giving up my life was not going to save him and I had zero odds of getting him off the road. If I stayed there maybe I would be dead in a few short minutes. If I stood to try and drag Schultz, I was too easy a target for Charlie (nickname for the enemy). So, I jumped up and sprinted back off the road, hearing some AW rounds whizzing by me and dived behind the hedgerow. To my front was chaos as I heard *TAT-TAT-TAT… TAT-TAT,* pause, *TAT-TAT-TAT* again and again coming from the APC's and the crack of M16's on full auto chewing up real estate to my side. I picked my way back toward the skirmish line we had set up, dodging bullets while I worked away from the heavy concentration of fire. The firing was so intense that moving any of the troops or APC's forward was not going to happen. I heard a continuous stream of RPG explosions, *WHOOSH* then *a sharp BOOOM*, as shrapnel rained down and around me and the steady *CRACK, CRACK* of AK-47's. As I made my way past an APC, the tank commander was firing his .50 cal machine toward the hedgerow to his front as his driver was backing up in reverse. Another APC was burning, having taken a direct hit from an RPG. The track commander was dead. We seemed to be having trouble setting up a skirmish line to lay down a heavy base of fire at the enemy.

I desperately looked to find my platoon and give a situation report to one of the NCO's (non-commissioned officers). Overhead I heard the unmistakable sound of incoming artillery fire as it whistled overhead, followed by a second of silence before hearing that familiar *KAARUMP, KAARUMP* of exploding HE. I glanced over my shoulder and saw a plume of dark gray smoke and dust rising above the jungle canopy. Staying as low as I could as I maneuvered myself across the battlefield, I moved from one location to the next trying to spot someone I could recognize from my platoon. Another volley of artillery fire from a nearby FSB whistled overhead again toward the enemy positions with deadly accuracy.

Looking around, I spot SGT Hugh Bishop, "Sarge, Sarge, I need help to retrieve Schultz. We were together up front of the tracks. He got hit and I couldn't move him. We need to get up there now. He's right out in the open."

Sarge told me it isn't possible. "There's no way we can move up there with all this incoming fire and we're calling in fire support. Is he still alive?" he asked.

"I'm not sure," I told him as my mind goes back over what I believed to be true, that he was dead. "He never moved once he was hit and I couldn't feel a pulse nor see where he was hit."

"We can't risk it. Did you get hit?" he asked, when he noticed that I was bleeding from my wound. He summoned a medic over. I was never questioned why Schultz and I abandoned the MG crew.

I was ordered to remain there and await a Dustoff, call sign for a helicopter ambulance. The medic affixed a white tag to my shirt, identifying my wound, I would guess. While sitting there, the artillery began to pour in on the tree line across the road. Soon, we had gunships from A Troop, 3/17th providing support. Not long after that, a request was put in for TAC Air, close ground air support from Tan Son Nhut airbase. The Air Force sent out a pair of F-100's from the 3rd Tactical Fighter Wing and began dropping napalm. They were followed by another flight which dropped 250 lb contact bombs. They created a column of gray dust and smoke as they struck the ground. We could feel the concussion from the short distance we were from the airstrikes.

The call for a Dustoff went out to pick up our wounded. They were late arriving. Within a half hour or so, my squad leader who was also wounded, SGT John Spoores, and I were picked up by MG (Major General) F.K. Mearns, the 25th ID commander in his C&C, (command & control) Huey, who was monitoring the net traffic and realized that he could assist in performing medical evacuations.

Apparently this dogfight was big enough for him to show up and see what was going on. Right after I climbed aboard his C&C Huey, he motioned for me to put on a headset.

"Can you hear me son?" he asked.

"Yes sir," I replied.

"Good, tell me what the situation is down there?" the General asked.

"Well sir, we had come across a few dead bodies as we started to sweep the jungle. A short time later one of the track commander's spots movement to his front." I continued on, "He didn't want to advance so Schultz and I moved ahead of the track and the rest of the formation onto the oxcart trail and all hell broke loose when the enemy opened up on us. There were VC everywhere in bunkers. Schultz was hit in the exchange and we need to get him out of there."

The General keyed his mic again, "What else can you tell me?"

"The tracks answered back with their .50 cal's and lots of small arms fire from Charlie and Bravo companies. Everyone was firing back hard at the enemy, but we were taking some casualties," I said. "The gooks were throwing lots of RPG's at us."

The General said to me, "Thanks, we'll get you boys to the aid station."

MG Mearns flew us to Cu Chi to the 12th Evacuation Hospital for medical treatment. Spoores and I saluted the General as we got off the chopper and he lifted off, headed back to the firefight.

Cu Chi was the 25th ID basecamp and home to the 7th Surgical Hospital and the 12th Evacuation Hospital. They arrived in Vietnam in 1966 and served and treated men from the 25th ID and other units until they returned home in the fall of 1970. There was a 60 bed hospital there with four operating rooms. We were met on the helipad by the medical staff and were walked into the treatment room where our wounds were to be cleansed and dressed.

"Where are you hit, soldier?" the doctor asked.

"I got it in the right arm. It feels ok," I replied.

After poking around my wound, the surgeon told me he had decided not to probe any further and try to remove the bullet as it may cause more harm than leaving it in.

"You'll be fine in a week or so. I'll clean this up as best I can and send you on your way." He trimmed the loose flesh from around the wound, flooded it with peroxide several times, and watched the blood lather up like putting vinegar on baking soda. It didn't hurt and did a good job of flooding out particles of dirt from the wound.

The doctor and surgical nurse left the wound open and didn't stitch it closed, then applied a gauze dressing. "Check in at the aid station tomorrow and have them clean it again, remove and replace the dressing," Doc said.

"OK, thanks, Doc," I replied, then headed outside for some fresh air. I fumbled around trying to unbutton my fatigue pocket so I could get to a pack of cigarettes. I removed the pack, opened it and removed and lit up a cigarette as I waited for Spoores to be treated as well. When I first arrived in Nam, I tried smoking regular cigarettes, like Winston or Marlboro, but they were too harsh on the throat, so I switched to Salem menthols.

PFC Barry Price, who arrived in country with me was also wounded and was evacuated after I was picked up. He had blood covering his face. Once he got to 12th Evac, they began to wash the blood off. His wounds were superficial shrapnel wounds and looked worse than they were. Head wounds always look more imposing than they are when we bleed just because of the heat and surrounding air temperature. They couldn't find any wound to stitch up and he was treated and released.

SGT Spoores, Price, and I headed over to the airfield to see if we could catch a *short hop* up to Dau Tieng. No problem, there was an aircraft leaving in about an hour. It was a quick 10 minute flight up there.

Our C-123 Fairchild Provider, a two engine military transport aircraft designed for short runway takeoffs, banked sharply and dived down through the sky toward the east end of the runway, attempting to avoid ground fire from the VC who were constantly roaming the rubber plantation. The runway at Camp Rainier was made of hard packed earth covered with crushed red pumice. The plane touched down and the engine thrusters were set in reverse. We taxied to the end of the runway and debarked from the plane. The camp was situated right next to the village of Dau Tieng and the southern side was nestled in a grove of rubber trees. This entire area belonged to the Michelin Rubber Plantation and was started up by the French who colonized this area back in the 1800's. Some of the plantation buildings had been repurposed for the Army's use. There was a large two story building that became brigade headquarters. Other buildings within the basecamp housed the First Aid station, the barber shop, mess hall, EM and Officers Club, PX, barracks for the companies, and the laundry.

After our flight to Dau Tieng, we waited for word of the battle. Over the radio we heard that Dr. Martin Luther King had been assassinated back home in the *world* as we called it. April 4th would be remembered for many of us in more ways than one. Checking in at the company HQ, we heard that air support and heavy use of artillery was being used and that we were still in contact.

The next day, the 5th, we heard more news and that Knapp was killed too. Schultz's body was recovered late that afternoon, along with SSG Tommy Knapp, our platoon sergeant, and two other KIA's from the 3/17th Cav, SSG Richard Call and SGT Heinrich Gerstheimer, when their track was hit by an RPG. Bravo Co lost two men in the battle, SGT Lawrence Osborne and SP4 David Rosenberger. The units swept the area after the NVA pulled out, finding four KIA's and blood trails for another four possible dead. In the afternoon, the task force was airlifted to the west where the three companies set up in separate laager sites for the night. During that day, A Troop 3/17th Cav as well as a team from F Co, 51st Inf Long Range Reconnaissance Patrol or LRRP's (pronounced lurps), as I learned to call them later, spotted a lot of activity on the ground. The decision was made to go and investigate. A Troop spotted 25 fresh gravesites west of our contact area. Could it be that the NVA removed their dead and packed them several miles before they buried them? They were known to do that sort of thing. Maybe we got more than the eight credited kills from that engagement. I also remembered this action being reported on TV back in the world, and that we had been engaging an NVA battalion and were credited with killing a hundred NVA, but that statement just wasn't true.

For the night of the 5th, the men hunkered down where they were, weary from two days of tension and bloodshed. The units swept the area on the 6th, relocated further west and on the 7th once again they reengaged the enemy. We lost several more KIA's in the process, but got our revenge. The battalion remained in constant contact until the 9th of April. After that action, the 2/12th was returned to Dau Tieng by helicopter and the 199th LIB was released from the control of the 25th ID.

In my first firefight, I was wounded. I thought to myself, this could be a long year. In the battle our losses were eight killed and too many wounded to count.

According to PFC Elmer Lightner who was in 1st Platoon, "I'm not sure just how many days the battle lasted but I know there was a lot of artillery and air support before it was over. After it was over we walked to a basecamp that had been set up with a battery of 105's and I know A Co 2/12th Inf was guarding the guns. I'm sure of that because a buddy of mine was with A Co and I looked for him when we got there only to find out he had been hit in a mortar attack a couple days earlier and dusted off. I do know another unit came through behind us and found a bunch of weapons and stuff and I believe it was either the 198th or the 199th, but I couldn't say which."

While the company was still away conducting operations, I had a chance to explore the basecamp. Camp Rainier was built by the 2/12th *White Warriors* and elements of the 2/22nd (M) Mech known as *Triple Deuce* and the *Regulars*, the 3/22nd Infantry. The basecamp could house several thousand men. The outside perimeter of the camp was guarded by rows of concertina wire, Claymore mines and bunkers were built about every 100 feet. They were equipped with firing ports and could hold 4-6 men or more. There was a berm that also encircled the camp and was located right behind the concertina wire. The bunkers were manned every night and a watchful eye was kept on the unfriendly terrain that butted up against the basecamp. We were also protected by a battery of 105mm howitzers. In the middle of the camp was the runway and at the western end, there were revetments for the helicopters that were based there occasionally.

It was a short walk from our company area to the Brigade HQ near the airfield, an old two story building likely used as an office building by the French to get to the Aid Station. I walked thru the rubber trees of our bivouac area. Our sleeping quarters were made of half walls constructed with wooden planks about four feet tall. The rest of the side walls were covered with screen and we had a canvass roof. Each soldier had a cot and a metal locker to store personal effects. The floors were wooden and elevated off the jungle floor. Most lockers were decorated with centerfolds from Playboy magazine.

Twenty paces from our hooch (building) was the outhouse, set back near the perimeter. It was a four-holer and got plenty of use. It had taken me only a few days to develop dysentery and get the runs. When you felt a bowel movement coming, there was no wasting time for fear of depositing that mess in my fatigues.

I arrived at the Aid station in the morning, located on the bottom floor of a two story building. It was painted a soft yellow and appeared to be covered in stucco. I entered through an open door and immediately was asked "What do you need?" by Doc Wade Lasister, who was in charge of the aid station.

"I got hit yesterday and was treated at 12th Evac. They said I should check in here with you to check out my wound," I replied.

Doc looked at me then said, "Take a seat and let me see what you've got." He removed the bandage then examined the wound. "Not bad, I'll change the bandage after I clean this wound up some more." Lasister continued, "This might sting just a little bit" as he poured peroxide on my wound.

I watched the liquid fizz and bubble on the open wound. Lasister patted the wound with a dry swab, then applied a new dressing and told me to come back the next day so he could repeat the process.

"You're on light duty until further notice," he said and handed me a medical note for the First Sergeant. "Give this note to Top and I'll see you tomorrow. You'll probably be on light duty for 7-10 days, but we'll see how quickly you heal up."

While I was there, I asked for something to settle my stomach and for the runs. I remained on light duty, along with Spoores. We spent the time writing letters home, talking with the other guys on restriction, and hanging out in our hooch. About eight days later, we were given a clearance to return to full duty.

As I sat around Dau Tieng on the mend, I had time to look back on the past year or so that got me to where I was now, sitting 8,000 miles from home.

2. DRAFTED — CIVILIAN LIFE ENDS

FALL OF 1967

My career with Pacific Bell was abruptly interrupted in the middle of the summer of 1967. I was enjoying life living in a two bedroom apartment in Rincon Valley, which lies several miles to the east of Santa Rosa, California. I grew up in this valley, having spent my childhood over on Hanson Drive, about a half mile away where my father had a house build in 1949. I was living in the heart of wine country in Sonoma County. I was sharing my apartment with my girlfriend, Judy who had just moved in with me. She was attending Redwood Beauty College learning to be a hairdresser. Our relationship was just warming up at the time. Let me restate that, it went from warm to hot in a flash, and toss in a beer or two and she was sizzling. Having graduated from Santa Rosa Junior College, I knew that my draft status classification would change from 2A to 1A and that gave me some angst. The news on TV and in the papers was all about the Vietnam War and that President Lyndon Johnson was escalating matters as part of his re-election bid. It seemed only a matter of time before I was going to get some very unpleasant news from the Draft Board.

The prospect of being involved in a war did not fit into my vision of my future. Somehow, some way, I just knew that I was not going to escape from being dragged into this world crisis.

I tried to forget about the war news and was making the best of my newfound freedom of living away from home. Early on, it was a struggle. I was just making enough money to pay my rent, utilities, and some food. I was surviving on peanut butter and jelly sandwiches half the time. I walked to work since I did not have enough money to buy a car which would change in a few months. Moving from Cloverdale, where my parents were living, back to Santa Rosa, got me hooked back up with my childhood friend, Bill Hougen. I was making new friends, going to music concerts, and enjoying my new life style. This dream state did not last long. Sometime in mid-July, the Selection Board had other plans and my number was selected, unbeknown to me. They were too kind and sent me a wonderful letter that landed in my mailbox.

"Dear Sir, we take this opportunity to inform you that you have been selected to join the U.S Army, etc., etc. Your date to report is September 6th at the Oakland Army base," blah, blah, and blah.

Wow, this was just a great how do you do letter. I did my best to get what affairs I had in order. I was in somewhat of a quandary as to what I was going to say or do in regards to my girlfriend, Judy. I'm not sure which brain is telling me what to do. My last weekend flew by and before I went to bed, I set the alarm for 5am. My sister gave me a ride and dropped me off at the bus station, where we said our good-byes. It was a lonely walk getting onto that 6am Greyhound bus. On Monday, September 6, 1967, a month before my 20th birthday, I was sworn in at the Oakland Army Base, California.

That previous Friday, my last as a civilian, I was in court attempting to challenge a speeding ticket I had gotten.

"Your honor, let me explain," I started out to say, but the judge wasn't hearing any of my whining and excuses.

"That's what they all say, and I've heard it all before," he replied sarcastically.

To make matters worse, when I told him, "Your Honor, I have been drafted and I am reporting in to the Army on Monday, and I have no money."

"Fine! Bailiff, book him. He can sit in jail until Monday," he retorted.

What the F****? I say to myself as they removed me from the courtroom. After they had chosen the right image out of several snapshots they had taken of me holding up that sign with my name and criminal number on it, I was allowed to call Judy and told her to get down to the jail and bail me out. Fortunately, she had the money to do that. My last 48 hours as a civilian would go quickly.

Now prior to all of this, I had given some serious thought to my relationship with Judy and we talked about our future. I decided to propose to her (minus any ring) and she accepted. I did not share this bit of news with my family for some reason. Along with this proposal, I left her my worldly goods and my apartment, but my '63 Corvette Stingray hardtop convertible, the keys to that were given to my youngest sister, Suzie.

From the Port of Oakland, myself and the rest of the new recruits were bused to Travis AFB near Fairfield, California and loaded on a commercial jet bound for SEATAC, Seattle Tacoma airport. From there I was transported to Fort Lewis, Washington for BCT (basic combat training). The day we arrived at the base, we were given fatigues and assigned to barracks. After mess call at 1800 hrs, we settled into our bunks for the night. That vision of a good night's sleep quickly ended when we were rousted from our beds around 0130 hrs and told to get dressed because the Army thought it was a good time to give us our psych exams. I apparently failed mine and was not offered any special training; they thought that my answers about liking to fish and hunt and camp meant I would like the infantry. Stupid me! One of the things about Fort Lewis is the fact that it rains there a lot. It seemed for every day of sunshine we got rain, cold wet rain. I think the funniest thing about basic training was the day right after we got there that we were marched down to the barbershop for our haircuts. It was comical to watch recruits sit down in the barber chair and the barber would ask them how they wanted their haircut, then proceeded to give them a number one buzz cut. All those curly locks, short and long, quickly ended up on the barber shop floor. I don't think we were there more than an hour and everybody in our platoon had a new look.

Our drill sergeants were veterans of the Vietnam War. I don't remember their names but I sure can remember their stature. We had one sergeant from the South who chewed tobacco and had served with the 4th ID. The other one was tall and skinny as a rail, black, with huge feet who served with the 25th ID. They were tough on us but fair. One thing we quickly learned and that was we never walk anywhere. I was happy to see basic training end.

My eight weeks of basic training at Fort Lewis quickly got my mind on track and around the idea that I no longer was allowed to be an individual, but a team player. We trained hard, running from point A to point B nonstop, rain or shine. I was only into my training for six weeks when I got a "Dear John" letter from Judy telling me our engagement was off and she was engaged to someone else. Under my breath, I said, "You're fu***ing kidding me?" This was all I needed right now, and there was nothing I could do or say about it. For weeks, I went to bed at night just thinking about that letter she had sent me and here I was, stuck under the total control of the U.S. Army. Whatever I wanted to do in response would just have to wait.

We had one guy in the platoon which we seemed to drag everywhere. He was physically weak and we really wondered how he got drafted in the first place. Well, we soon learned what that was all about. This older gentleman was singled out during company parade and given captain's bars on graduation day. BCT (basic combat training), 2nd Plt, E Co, 1st Bn, 1st Bde ended on November 10th and after a brief three hour visit at the parade grounds with my parents, who drove up for the graduation ceremony from their home in Cloverdale, California, I headed to my next duty station for AIT training. AIT stood for advanced individual training and is specific to your assigned duties in the military. Graduating from Basic at Fort Lewis led me to stepping onto a chartered Lockheed L-1649 Constellation, the old four engine turbo prop driven passenger plane. We left SEATAC (Seattle Tacoma) airport around 1700 hrs and landed at Alexandria, Louisiana, at midnight. After we debarked from the plane, we walked across the tarmac and stepped into buses which were parked at the airport for our one hour journey to Fort Polk, which is close to Lake Charles and Shreveport. I would spend the next three months trying to stay warm there.

FORT POLK, LOUISIANA,
HOME OF TIGERLAND AND MY AIT DUTY STATION

After taking my 30 day leave at home in February of 1968, I was mentally prepared for whatever the outcome would be for my deployment to South Vietnam. I would say from a training perspective, I could have been more prepared. Why do I bring this up now? Let me take you back to Fort Polk, Louisiana, in the fall of 1967. I, along with 30 something other guys were thrown into 3rd platoon, and was assigned to D Co, 4th Bn, 5th Training Brigade. We were a platoon consisting of U.S.'s. Everyone who enlisted into the Army had two metal identification tags with the soldier's service number stamped into them that started with R.A., regular army. Draftee's on the other hand, had ID tags starting with U.S., more commonly known as dog tags. The dog tags had your name, service number, blood type, and religious preference listed on them.

We started our training in our MOS (military occupation classification for what you did in the army). The first three combination of numbers and letters identified what you did, the next two digits indicated your rank. 11B, which stands for infantry and 10 which indicated either a PVT or Private (also displayed as E1 or E2). The soldier could also be a PFC (Private First Class or E3), all indicated by "10", in this case, an MOS of 11B10. Everyone had to be a PFC upon arrival in Vietnam unless you had been disciplined and reduced in rank for some infraction.

I never realized it would or could be so cold in the south. After being there for several weeks, I developed a cold which ultimately turned into pneumonia. I tried several times to go to sick call, only to be sent back to the company with some aspirin and cough drops. Sick call is the process for asking and seeing a medic or doctor. Finally, a week later, I was running a fever of 103 degrees and once again I went to sick call. This time I was sent from sick call at the infirmary to the hospital where I was put into a respiratory ward where the place was filled with trainees all suffering the same illness. We had to get out of bed at reveille which was 0600 hrs, and if we wanted to eat, make our way to the mess hall for all our meals. We could then return to our ward, but were not allowed to get back into bed and under the covers until 1700 hrs. I was there for several weeks.

With Christmas block leave approaching, I visited with the doctor and did my best to act as if I was fully recovered. Without being released, I had no chance of going home for a two week leave and would be stuck at the base. The Army called it a "block leave" meaning everyone, officers, staff, and trainees were granted leave and the base would be shut down for whatever period of time was deemed as block leave. As my exam with the M.D. progressed, I did my best to conceal any signs of illness. It was difficult trying not to cough, which would have been a signal that I was unfit to be released from the hospital, but I got lucky and with the help of my oldest sister Kathy paying my airfare, I got to go home for Christmas, which lasted from December 15th to January 2nd. I headed over to Alexandria to the airport the next day to fly home. I checked in with the airline, Trans Texas Airways, and waited for the announcement to board the plane. The voice over the loudspeaker said for all passengers who are traveling to Houston, that it was time to board and to head out to the plane.

What a surprise! It was a Douglas DC-3 tail dragger with the letters TTA painted on the tailfin, and the pilot was having a difficult time starting one of the engines. It was coughing and sputtering and just would not catch fire. A blue gray cloud of smoke was consuming the airplane and the passengers standing in line to board. Finally, the radial engine started to run and smooth out and with that the cloud of unburnt fuel and carbon monoxide began to dissipate. I climbed the ladder to get into the plane from the rear. I worked my way to my seat, by grabbing the backs of each row of seats, pulling myself along, finally reaching my assigned seat. The takeoff was OK but I was praying that we would get to Houston without crashing. The pilot seemed like he couldn't gain any altitude and we hugged the tree line the whole flight, finally landing and arriving safely about an hour later. At the Houston airport, I found my way to my gate for the next leg of my flight. The PA announced that boarding started for my American Airlines flight. I slept in the back of the plane since there were vacant seats. It was a good flight and the stewardess took good care of me along the way.

Arriving back at home again in Santa Rosa, California, it was good to see my Mom and Dad and my three sisters. After a day or two of visiting, I jumped in my Corvette and headed north to Miranda, a three hour drive north to confront Judy, my ex-fiancée face to face. I called her to let her know I was headed her way. It was a chilling reception that I received and she had no rational answers to my questions other than to say that she was moving on, dumping my ass in the process. The three hour return trip had me wiping more than a few tears off my cheek bones. It didn't help to listen to the radio and hear Gary Puckett and the Union Gap playing *"Young Girl"*. The chapter of "Judy" was closed for good, although not forgotten about as I continued to dwell over it for the next few months.

On my return trip back to Fort Polk, Louisiana via Houston, I learn that Trans Texas Airways cancelled my flight due to bad weather and bused us from Houston back to the base. I had enough of TTA after one trip. Someone coined the nickname of Tree Top Airlines and after my first flight with them that certainly held true.

Arriving back at Fort Polk, I was in for a surprise. Because I missed more than two weeks of training, I was removed from my training company and was told that I was being recycled and was to be reassigned to a new company in order to make up for the two missed weeks. Another kid, John Buetler from Grass Valley, California was in the same predicament. So, we hung out together awaiting our fate. Finally, a new training cycle was started, and we got our orders to join Echo Company. Since we had already completed three weeks of training, the C.O. and Top Sergeant had us go over to the motor pool and take a test to get a military driver's license. Upon completing those orders, Buetler and I began each day checking in with Top then going over to the motor pool and picking up the vehicles the company needed for the day. We usually got a Willy's Jeep and M37 Dodge 3/4 ton truck, but on rare occasions, we drove a M35 Deuce and a Half, a 2 ½ ton military truck. This was an all-wheel drive with four over and under gears with duel axels in the back and a canvas covered bed. We hung with the *staff* during this time while the recruits (troops) went through their formal training. We thought once the company got into the fourth week of training, our mission as company drivers would end. That was not to be.

With the exception of weapons qualifications with the M16, M60, and hand grenades, and a side junket to S.E.E. (survival, escape and evasion) training; we continued to be the company drivers. I'm not sure what the rest of the company thought about us two. The time the company had to go out on an overnight bivouac it was freezing cold with rain. John and I were nestled down next to the gas heater in the cadre tent, cozy as two peas in a pod.

We had a free run of the post whenever the company was not in formal training. Fort Polk had two parts, North and South post. When we knew we could get away with it, we would drive down to South post to see what was going on. Sometimes we would stop at the PX or stop at the cafeteria and pick up a hamburger, coke, and fries, then we would jump back in the Jeep and head back to the North post and turn the Jeep in at the motor pool. Fort Polk was built in the rolling hills of Louisiana covered in sand and pine trees. This was a perfect spot for training and preparing troops for war. Buetler and I continued as company drivers until graduation during the first week of February and afterward we were sent home with our orders in hand but knowing we had a 30-day leave prior to being sent overseas.

I spent my time at home visiting with friends, family, and hanging out with my good friend Berry Blackmon. Berry and his brother Rick, lived around the corner from me and we used to spend a lot of time together playing street sports, baseball and football, mostly with the rest of the neighborhood kids. In the 60's, cruising was the *in thing*. Whether you borrowed your dad's car or had one of your own, it got washed and waxed before you took it out downtown where every town had a pattern or loop that the kids drove. In Cloverdale, the route took you through the center of town on Main Street to the north end where the turnaround spot was Foster's Freeze, a burger stand, then back through town to the south end to the Owl Café. Foster's was the place to be and I would usually buy a cherry or vanilla coke to sip on as I cruised the strip.

I would see Patty Pressley driving her dad's big white Cadillac all the time. We would wave at each other as we passed by, but had never really stopped to talk. Patty, as it turns out, was the former fiancée of my future brother in law to be, Jim Miles, who lived in Booneville, a small logging and agricultural town about 30 miles from Cloverdale, a town three times the size with a population of 3300 residents. He would eventually marry my youngest sister Susie after I came back from Vietnam. Anyway, this routine soon gave me motive to stop Patty and find out what the story was about her because I never saw her with anyone. I wouldn't call her my girlfriend at this stage, just a friend, and we enjoyed each other's company along with a bit of chitchatting.

My parents moved us from Santa Rosa, to Cloverdale in 1959, on my birthday. I was in eighth grade at the time and did not adjust to the town and make friends easily. It had a four-year high school with a pretty good sports program. They won a number of small school state championships with their basketball teams. I was not large for my size then and because of that, lacked the confidence to try out for any sports. When I graduated in 1964, I weighed about 145 lbs. Two days before our graduation, three members from my Cloverdale High School were killed in a head-on collision with an eighteen wheel truck. Two of the dead were my classmates, Ricky Stevenson and Chris Bilbro. The third kid was Charles Greppe, a junior.

Our graduation was delayed for a few days. This put a dark cloud over the town for a long time. But, there were brighter moments living there, like swimming in Sulphur Creek or the Russian River that was just east of the town and there were plenty of places to bike or wander around in the woods hiking. My time at home quickly passed and before I left, Patty she said that she would do her best to write to me while I was in Vietnam. As I was leaving, I told her to go on with her life, date, or do whatever while I was gone. I told her there was no way of knowing if I'll be coming home or not. In all this short time, I can't remember if I had even kissed her...

3. LEADING UP TO MY ARRIVAL IN VIETNAM

JANUARY 31ST TO MARCH 25, 1968

Background: South Vietnam is divided up into four military zones, I Corp at the top of the country and butts up against the DMZ, demilitarized zone, which separates North from South Vietnam. II Corp area is called the Highlands, the narrow shape of the country, followed by III Corp, and home to the 25th Infantry Division, the area I would ultimately be assigned to once I reached Vietnam. It contains a major city and the capital, Saigon, and the smaller towns of Cu Chi, Dau Tieng, Tay Ninh, Trang Bang, Duc Hoa, and Hoc Mon, among others. These towns contain 200-500 residents. The Mekong Delta was referred to as IV Corp and was the responsibility of the 9th Infantry Division which arrived in December of 1966. There were other divisions also assigned within these regions, including the Marine Corps, which I will not detail here.

The split of the country at the 17th Parallel and referred to as the DMZ was established at the Geneva Conference in 1954. The documents are referred to as the *Geneva Accords*. After the French had lost to Ho Chi Minh's fighting forces at Dien Bien Phu in March of 1954, Ho Chi Minh wanted the country to be liberated from the French and recognized as the Democratic Republic of Vietnam for a second time. The first being after the Japanese occupation forces were defeated at the end of WWII. He did not get this recognition then from the international community and they seemed unwilling after the French defeat as well, the second time. The Soviet Union, the United States, France, the United Kingdom, and the Peoples Republic of China participated in a conference held in Geneva, Switzerland to resolve outstanding issues resulting from the Korean War and to address restoring peace in Indochina after this action. The accords stated that the north would be governed by the Viet Minh (Ho Chi Minh's party) and in the south by the State of Vietnam.

General elections were scheduled to take place in 1956 to establish a unified Vietnamese state. The State of Vietnam never agreed nor signed the Accords and at the same time, the North never honored any of the provisions of the documents they had agreed to either. The elections never took place once the State of Vietnam declared its independence in 1955 and was recognized as the Republic of Vietnam by the International community. From that moment on, North Vietnam vowed to repatriate the country and immediately began to prepare for war.

While I was trying to enjoy my time at home before I was to be shipped overseas, all hell was breaking loose in Vietnam. This chapter addresses what the Battalion was going through during the TET Offensive in the months of February and March, just prior to my arrival in country.

In two weeks of continuous contact in January and February, elements of six 25th ID battalions had killed more than 400 Viet Cong during the fighting in the Cu Chi area. Battles raged along Highway 1 from the outskirts of Saigon to the north of the division's basecamp at Cu Chi. Heavy fighting also broke out during the Tet period from the Hobo Woods to Duc Hoa in the northern and southern extremes of Hau Nghia Province.

Soon after the Viet Cong shattered their declared Tet truce on January 30, 1968, the 1/27th & 2/27th *Wolfhounds* were airlifted into the Saigon area to reinforce American units defending the Capital. When it became apparent that more troops would be needed to handle the string of coordinated attacks along Hau Nghia Province's stretch of Highway 1, units were dispatched from the 3rd Brigade, 25th ID on February 1st.

Both the 3/22nd and the 2/12th came in heavy contact within hours of arriving under 2nd Bde control at Cu Chi. The 1st Brigade's 4/23rd (M) Inf already under the operational control of the 2nd Brigade on a land clearing operation in the Hobo Woods, also saw heavy action in the Viet Cong Tet offensive.

The 2/27th was airlifted into Tan Son Nhut Air Base where 3/4 Cav (3rd Squadron, 4th Cavalry) and their M48 Patton *Big Boy* tanks were repulsing a massive enemy assault. The *Wolfhound* battalion set up a base nearby and began to battle enemy units poised to strike at the Tan Son Nhut military complex and at the Capital. Also on January 31st, an ambush patrol from the 1/27th killed 15 Viet Cong and captured a 75mm recoilless rifle. Reinforcements from the battalion's forward base at Duc Hoa killed 22 more and captured a second 75mm recoilless rifle.

Three companies made helicopter assaults into the Saigon suburb of Hoc Mon, which the Viet Cong had overrun the night before. The American force immediately began clearing operations. In the early stages of the fighting around Cu Chi, the 3/22nd and the 4/23rd (M) bore the brunt of the action. Within days, however, the 2/12th also became fixed in a continuous struggle to push entrenched Viet Cong from two villages to the east of the Cu Chi basecamp.

Early on the morning of February 1st, the ARVN Cu Chi subsector reported it was under attack by an estimated Viet Cong battalion. The reconnaissance platoon of the 4/23rd (M) led a company of the 3/22nd in a daring charge through heavy enemy automatic weapons and recoilless rifle fire to relieve the burning sub-sector compound. At daylight, three more companies of the 3/22nd joined the fight in a three-pronged attack into the village of Cu Chi. Street-fighting raged for five hours until the enemy force abandoned the village. The enemy unit, identified as the 1st BN, MR IV Main Force Regiment, suffered at least 12 killed, five weapons lost, and 21 personnel detained.

The fighting in Tan Phu Trung and Ap Cho, neighboring villages along Highway 1, less than 10 kilometers from Cu Chi, began January 31st when the 3/22nd moved in to clear two enemy companies blocking the road. In almost five hours of continuous contact, the infantrymen killed 17 enemies. The Viet Cong unit, believed to be from the 272nd Regiment, appeared to withdraw from the village. Two days later, however, fighting again erupted in the two communities, this time as a truck convoy attempted to pass through from Saigon to Cu Chi. While the convoy waited three kilometers to the south, a company from the 4/23rd (M) and elements of the 3/22nd fought their way into the town.

Although the 3rd Bde had borne the majority of the action in Ap Cho and Tan Phu Trung, elements of the 3/4 Cav; Alpha Troop, 2-34th Armor; and the 4/23rd (M) had assisted the 3/22nd in its attempts to drive through the enemy fortifications. So far, at least 106 Viet Cong had died in the American assaults on their fortifications, lost six individual and two crew served weapons, and several personnel detained. The 2/12th killed at least 115 Viet Cong in the battle to clear Highway 8A at Tan Hoa, and in action between there and An Phu Trung, six kilometers to the south. Also faced with Viet Cong entrenched in reinforced concrete bunkers, the 3rd Bde unit fought against elements of an estimated two battalions of the 272nd Viet

Cong Regiment. Although the American unit had faced sporadic contact shortly after it airlifted into the Cu Chi area, its first significant contact came on February 5th, when it killed 33 Viet Cong who had dug in at the hamlet of Phuoc Hung. The following day, the infantrymen moved north in an attempt to drive through the village of Tan Hoa. Like the action along Highway 1, the fighting had raged ever since.

On February 6th, a company of the battalion killed 22 Viet Cong who had opened fire with small arms, machine guns, and rockets. Artillery fire from Cu Chi and armed helicopters supported the troops in their assault. Air Force fighter bombers dumped thousands of pounds of explosives onto the enemy fortifications during the eight day battle. Although the U.S. troops several times penetrated the enemy defenses and captured several automatic and crew served weapons, the Viet Cong force had continued to resist with heavy fire, all attempts to break through the town.

In other actions throughout the province, the 4/23rd (M) killed 24 Viet Cong and captured one RPG-2 rocket launcher in a five-hour fight in the Hobo Woods. Throughout the two week period, tactical air strikes accounted for 35 Viet Cong killed, artillery 30, and helicopter gunships, 25 enemy killed. Two kilometers to the east, two companies of the 2/27th were locked in battle with a large enemy force. Fighting until dark, the infantrymen killed 102 enemy, and were supported by tactical air strikes, artillery, and gunships.

The first three months of the year, the NVA and VC tried unsuccessfully to root us from fire support bases Buell, Rawlins, and Tay Ninh and Dau Tieng basecamps. They threw a ton of men and equipment at us and in each case, he was dealt a severe defeat. Every time they mounted a ground assault, we chewed him up and spit his remains on the battlefield. The enemy's losses were staggering compared to ours. In the end, they were forced to remove what main body of troops they had left and sought shelter across the border in Cambodia or moved to some extreme safe havens in War Zone C. The intelligence we had collected suggested that what remained behind were smaller local units of VC who would coordinate the staging of food supplies and weapons for use later on when the main forces re-entered the country after they had gathered new replacement troops from the north.

4. DAU TIENG AND CAMP RAINIER

I spent the last remaining few days saying my goodbyes to all my friends and family, Mom and Dad. My favorite uncle, Richard, my Dad's brother who lives in Oakland, came up to Cloverdale to see me one last time. When I was younger, he would always give me a few dollars for my pocket when the family went to Oakland to visit with him and my grandmother. His favorite meal on Sunday's was fried chicken and ice cream for dessert.

My travel orders said I was to report into the Oakland Army Base and debarkation terminal on March 10th where I spent several days waiting for my next set of orders. On March 12th, I was handed my manifest sheet, climbed on board a bus for the 30-minute trip to Travis AFB at Fairfield, where I boarded a Boeing 707 chartered by Tiger Airlines around 1100 hours. Take off was smooth and everyone settled in for the long journey across the Pacific Ocean. A layover stop in Honolulu lasted several hours and at Clark Air Force Base, Manila, Philippines where we had to refuel, then continued on where we arrived at Bien Hoa AFB around 1400 hours on March 13th, some 15 hours later. As the doors opened on the aircraft, I got a strong whiff of the country I was about to live in for the next year. It didn't exactly smell like a garbage dump, but it wasn't roses and citrus flowers either.

With the air hot, muggy, and heavy with humidity, we were transported over to the 90th Replacement Center known as Camp Alpha in Long Binh, to await our unit assignment. I ended up staying there about five days.

To keep us busy, we were given fatigues to wear, and sent out to pass the time filling sand bags. These were woven polypropylene and dark green in color. The sun was bright and the heat matched its intensity and the sweat was abundant on everyone's brow as we shoveled sand for hours on end. Some of us, obviously new to this foreign country, did not realize how quickly you could get sunburned. I was not one of them. Some of the guys took off their shirts and within hours, looked like cooked lobsters. We spent two or three days here awaiting orders and assignments to our field units. My original orders had me reporting to the 1st ID, or so I thought, but that was to change. I finally was ordered to report to Cu Chi on the 18th where the 25th ID HQ and basecamp were located. I hopped aboard a C-7A Caribou, a small high winged, twin engine aircraft, and off we went, flying in a westerly direction. Upon arriving, we milled around and waited a few more days before we were sent to mandatory orientation training that took three days to complete. This training gave us some perspective on what combat would be like, pointing out the hazards of booby traps, weapons used, details of terrain and our enemy, and gave us some idea of what to expect from the local peasants and ARVN troops.

After graduating, or more like, just completing this orientation class, I received further orders to join up with C Co, 2nd battalion, 12th Infantry "White Warriors" Regiment, which was in the process of relocating from around Hoc Mon, Tan Hoa, and Cu Chi area to Camp Rainier, Dau Tieng. LTC Charles Bauer was the battalion commander, having just replaced LTC R. Dean Tice the day I arrived in Vietnam. Tice eventually retired as a LTG (3 star general). The southern side of the camp was covered with rubber trees and that was also the bivouac area of the Warriors. We were lucky because the 2/22nd and 3/22nd had the northern area which was void of trees and shade cover. Our basecamp sat right next to the village of Dau Tieng. It was a VC village through and through. It was off limits to all personnel. The only time we went there or through there was on a tactical operation.

This nice little community is about 50 miles northwest of Saigon and sits between the Ben Cui rubber plantation on its left, separated by the Saigon River and to its right is the Michelin Rubber Plantation, both developed by the French who controlled this country, once call French Indochina, for decades. To the north is the Razorback mountain range and Pine Ridge. I was supposed to fly up there on March 26th, but due to a lack of flyable aircraft, I didn't arrive until March 27th. So, that next morning I moseyed over to the airport and boarded a C-123 transport plane for the short hop up to Dau Tieng. Landing the plane is a term I would not use. It was more like we dove down to the airstrip to avoid any sniper fire from the surrounding canopy of the Michelin Rubber Plantation.

I reported in at company headquarters and was told I would be joining up with 3rd platoon and 4th squad as an ammo bearer. The next day, all the new replacements were sent over to the armory and I was given a brand new M16A1 rifle and fifteen magazines, then issued my field equipment of four canteens, equipment harness, first aid pouch, two ammo pouches and belt, gas mask, back pack frame, poncho and liner and some basic clothing. It was here that I met SGT John Spoores, my squad leader from Indiana, and PFC David Schultz from Kent, Washington. David was a big kid with broad shoulders and a happy smile.

Also, arriving about the same week was William Hill, a black kid from New York with striking features and a big smile; Barry Price, part Cherokee Indian, from Loretto, Kentucky; Sidney Fowler from Georgia, William Potts and Mike Rinkle, who joined me in 3rd platoon. Reporting to first platoon was a red headed kid, Billy Zimmerman, who became the platoon leader's RTO, Craig Schoonderwoerd from Southern California, Jessie Anderson, also a black kid from the east coast, Leon Barnett, Elmer Lightner, from Pennsylvania, Hiram Marziano, from North Carolina, and others I'll name later. I would eventually gain a close relationship with most of these guys. All in all, the company added 40 new replacements during the month of March, which was somewhere between 35 to 45% of the company's strength.

I think as each of us settled into our new lives and job, we all had the same questions. We had had a few days to absorb a few stories about the company's recent return from Tan Hoa and Hoc Mon in the middle of the TET Offensive. They got pretty banged up in the process. What is combat like? Will I survive and return home, and in one piece? Will I do my job as I was trained? How long will I last under the stress and still be functional to my squad, my friends, and my platoon? We would know the answers to many of these questions in only a few short days.

We set about prepping our equipment, cleaning our rifles of Cosmoline, a wax-like petroleum based corrosion inhibitor that new rifles are subjected to for storage until needed, and loading fifteen magazines with 5.56mm ball ammunition. I decide to load mine with some tracer rounds. Matter of fact, every fifth round was a tracer. SGT Spoores told us to load only eighteen rounds into the magazines. He said before we lock a magazine into our rifle, to slap it against our thigh to loosen the rounds so they will be sure to feed properly into the chamber when firing.

Once we had our equipment needs taken care of and stored for ready access, it was time to take a few minutes out of the day to write home to a number of people, Mom and Dad and other members of the family and give them my contact information and address for the unit so I could receive mail. I penned some quick letters telling everyone where I was, how my trip to here went, and a brief description of my new surroundings. I quickly finished up that project, then headed off to the chow hall for dinner.

Note: After my first firefight, I reloaded my magazines and removed the tracers. I had learned an important rule of combat: "Tracers work both ways and I didn't want Charlie being able to find me so easily." The boys firing the M60 machine guns needed the tracer rounds to pinpoint their firepower, but it came with a cost and attracted a return fire of RPG's.

It took no more than the time I was in country to come down with dysentery. Every few hours, it was a sprint to the outhouse for a lengthy visit on the can. Over time, this went away, but it was tough on the system, getting adjusted to the local water. And to pass our time, we had an occasional drill to find shelter in the bunkers from the incoming 82mm mortar rounds the Viet Cong (VC or Charlie) threw our way.

There was a lot of controversy surrounding the Army's choice of rifles to replace the M14, a nine pound semi-automatic which fired a 7.62mm cartridge. They settled on the M16, manufactured by Colt, an adapted design of the ArmaLite AR-15. It began to earn a bad reputation early on when it was first used in combat. The rifle was prone to jam and this caused a few untimely deaths as soldiers frantically tried to clear their weapon of the jammed cartridge still hung in the firing chamber. They resorted to carrying cleaning rods into the field which they would use to ram down the barrel in hopes of knocking the spent casing out of the firing chamber.

I received a brand new modified M16A1 version, which had an added "forward assist" or button used to seat a round in the chamber if the return spring had not done so, updated black chrome bolt and firing chamber. The three prong muzzle flash suppressor which some guys used as a tool to break the bindings on C-ration cases, was replaced with an enclosed flash suppressor. It was billed as a self-cleaning rifle. The problem with the first issue of Colt M16's was directly the Army's fault. The weapon was designed to use IMR 8208M or IMR 4475 extruded powder. DuPont informed the Army that they could not mass produce this powder to the specifications demanded by the M16. Therefore Olin Mathieson Co provided a high-performance ball propellant. While the Olin Mathieson WC846 ball powder achieved the desired 3,300 ft per second muzzle velocity, it did not burn as clean as the extruded powder. This residue quickly built up in the rifle and was the root cause of the empty casings getting stuck in the firing chamber and the extractor failing to eject the shell. I never had my rifle jam once during my tour. I thought it was an excellent weapon.

Camp Rainier was named after the mountain in Washington and the area from which the 2/12th came from. The land in which the basecamp was situated used to be headquarters for the Michelin rubber plantation that was set up by the French when they occupied this area in the 1920s. These heavily forested areas provided excellent cover for the Viet Cong to hide and travel in. They were concealed from air surveillance and difficult to spot at ground level due the brush and grasses that grew under the canopy.

When I joined the battalion, they had just returned on March 22nd to Dau Tieng after having spent time operating around Cu Chi and Hoc Mon during TET. The 25th ID was winding down Operation Saratoga, an effort to maintain the security of the basecamps and keep open all MSR's (main supply routes) within the region. This operation, of course was interrupted by the Tet Offensive which kicked off on January 30th. The third week of March, the company was OPCON'd from the 3rd to 1st Brigade and was in a stand down mode in which they spent several days out of the field from combat operations and got to rest, get cleaned up and take showers and get clean fatigues, get some training which translates into *Lessons Learned* reviews of do's and don'ts during combat operations, and perform equipment repairs and cleaning.

The unit needed this first break from the action. The whole unit was whooped from exhaustion after seeing heavy combat action from the Tet Offensive counterattacks. The battalion took some casualties and many wounded during that five to six week period.

For the medical staff, it was the same situation; they needed to restock and to inventory their medical supplies. It was also a chance to check the equipment, get a set of clean clothes, and write a letter or two. One of the drugs used to treat jungle rot was called Potassium Permanganate. Not being a medic, I'm not sure if there were other treatments as well. Back to this particular drug that came in the form of a pill, it had a shelf life like most drugs.

There was an unnamed medic who was tasked to get rid of a whole bottle, or more, several hundred pills of this stuff because they had exceeded their shelf life. What to do with these pills? The normal treatment called for dissolving some in water, then soaking the feet in the solution. It had a side effect in that the water turned purple, thus, the feet end up with a violet tint to them. This medic looked around and was trying to solve his problem of dumping these pills.

About that time, a water truck stopped by the hooch he was staying in. Looking around, and seeing no one, he decided to dump his stash of pills into the water truck. What he failed to realize is that this water truck was filling the tanks for the shower stalls and the driver had just left the EM shower and was on his way to the Officer showers. Yikes! Who, of all people would you guess that was going to use the shower first? It was the battalion executive officer, name withheld. He was NOT happy when he turned on the shower and was coated with purple tinted shower water and then discovered he had a violet glow. Whether he was actually tinted from the shower or this just made for a good story, I'm not sure if it was true, or what kind of comments he received at officer's mess.

After the refresher training was over, the company did a few local sweeps in the immediate area of the basecamp and into the Michelin Rubber Plantation. There were thousands and thousands of acres of rubber trees with a high canopy which filtered the sunlight that beat down from above. Underneath all this was sparse vegetation that reached upward for that canopy of light above and blocked your view in every direction. That was not friendly turf we were walking in. My squad leader gave us new guys a quick schooling on what we should be on the lookout for when we were on patrol. Basically, follow the lead of the veterans and stay low.

Dau Tieng was fully equipped to support more than three battalions of men. All the buildings for the basecamp had been completed by mid-1967. We had barracks, officer and EM clubs, mess halls, and a PX exchange. We had other military buildings such as brigade headquarters and first aid station housed in old existing French buildings. Of course, who could forget the swimming pool that was built above ground, including a diving platform? We had a few other shops like the barber shop and laundry which were manned by the local Vietnamese villagers. One of the barbers was later shot dead while working his other profession as a VC guerrilla fighter.

There is a story about the swimming pool that I heard soon after I arrived and was also written about in the Tropic Lightning News, the 25th Infantry Division's publication about a year before I arrived and which came out each week with stories of the war, listings of who received medals, promotions, and general informative articles about the goings on.

MARCH 20, 1967

Tropic Lightning News

"The 3rd Bde of the 4th Infantry Division recently added splash to one of man's most unusual wars when it resurrected a stately 1930 French swimming pool at Dau Tieng (Camp Rainier), just crackling distance from Viet Cong rifle fire.

In a scene that would undoubtedly send Lloyd Bridges under for the third time, the combat hardened third brigade troops have taken to the new swimming pool almost as calmly as a Vietcong mortar attack. "It's unbelievable!" said one specialist. "I haven't had this much fun since I left the States. About the only thing we don't have are girls."

And few would dispute that this was the greatest morale booster to hit the camp since Armed Forces television brought Batman to the country late last year. When the Ivy men arrived in War Zone C, it didn't take them long to uncover and overrun the dilapidated swimming pool, originally built under the shade of the Michelin rubber plantation by the French in the 1930s. The pool was marred by 20 years of neglect and war. The pipe system was beyond use, debris contaminated the bottom, a grenade explosion had ripped out a chunk of cement, but the potential was definitely there.

The 2nd Battalion, 12th Infantry, getting a chance to show its versatility, moved in, cleaned up, then moved out, not before installing 10 showers below the sundeck, dumping 80 gallons of paint on the pool's walls and 275,000 gallons of water to fill the tank. The result was a fabulous 81-foot long, 31-foot wide and 12-foot deep pool and an opening day splash in that resembled a school of salmon in a goldfish bowl.

Since opening day, the pool has taken on a few added attractions and now features a little something for almost everyone. The high divers have a 4 meter tower, while a spring board constructed from a helicopter rotor wing has been installed for the less daring. A wading pool tapering down to 4 feet accommodates those who just want to relax and forget the bitter turmoil in War Zone C. The sundeck is equipped with a stereo unit, and management has available masks, swim fins, and sparkles.

The pool is managed by SP5 Ray Croft. He is responsible for the care and maintenance of the pool and equipment and for pool safety. He is assisted by SP4 Jim Fulton, Richard Webster, and CPL Ron McAtee. In addition to their other duties, the pool crew teaches daily swimming classes for the non-swimmers in the brigade. The pool is open daily from 8 AM to 7 PM. No one yet has suggested putting up night lights for a midnight swim. That might be just too much."

Note: When I returned to Vietnam in 2010, I returned to Dau Tieng, and the swimming pool was still there. I should have stopped the car and checked out the condition of the pool. But viewing the pool from the outside, it looked like it was still in decent condition. The structure and paint looked fairly new. I could not say whether the pool was still being used as intended. According to Bill Comeau, A Co, there was a Playboy Bunny painted on the bottom of the pool when it was recondition back in 1966.

In the company areas, there were red fifty-five gallon drums scattered everywhere for fire protection along the dirt streets. Strategically located between buildings were drinking facilities and down the street from our bivouac area was the community shower which consisted of a large aircraft pontoon suspended overhead for water storage. There was a wooden 'modesty' fence surrounding the shower itself.

I was always one who wanted a nice clean scrubbed body when I climbed into bed. Even today, I cannot go to bed without having a shower first. In 'Nam the convenience was not there, especially when we were out on operations. Washing would just have to wait until we got back to basecamp. After leaving Dau Tieng with the battalion, matters would get worse and most of us resorted to taking some kind of sponge bath. I would grab my steel pot and fill it with water and find a bar of soap and a makeshift wash cloth and get busy. With the rainy season, washing became a matter of just standing outside with a bar of soap, but I'm getting a bit ahead of myself with my story.

One of the largest concerns about life in a basecamp or fire support base is to control any outbreak of disease and to try and keep living conditions sanitary. When we had out used slit trenches for toilet needs, outhouses were built to accommodate the natural cycles of life. Most of these outhouses had three or four holes (or toilet seats). Under each hole was a tub made out of half of a 55-gallon drum, to catch the human waste.

Everyone at one time or another was assigned to the shit burning detail. That was not a pleasant task and for some it was viewed as punishment and for those who thought that, they were probably right. This duty required the removal of each tub from the back of the outhouse or latrine. Whoever was assigned this duty would then mix diesel fuel into the waste material and then set it on fire. Sometimes it was just your turn and other times it was administered as punishment. The unlucky person had to stir this mess with a stick periodically to keep it burning. The waste would be burned down to charcoal, then removed and buried for health safety. Everyone knew what was happening when they saw this large column of black smoke rising up above the perimeter of the camp. The stench from this, and just the process itself would make a person gag, but it had to be done daily.

I was just starting to find some routine to my new surroundings when SGT Spoores poked his head into our barracks and told the squad to "grab your gear and assemble on the helo flight line in 20 minutes." That was my first aerial helicopter flight involving the platoon. Spoores told us we were being flown southwest of Tay Ninh to grid map location XT2335 to secure a perimeter around a downed helicopter. I got a rush of exhilaration and adrenaline as we arrived at the helo pad area. There were two Hueys waiting for us to climb aboard as their engines and turbines whined as they provided thrust to the whirling rotor blades. Spoores motioned us to get in. The door gunner signaled the pilot that we were seated, then the pilot applied throttle to the engine and the blades began to shake and rattle the carriage of the chopper. He pulled back on the *Collective* to gain lift with his left hand and then as the chopper rose, he pushed forward on the *Cyclic* or stick with his right and the chopper slowly began to move forward with its nose slightly down.

The pilot kept the chopper in this configuration as he continued to build up ground speed then pushed forward on the cyclic and we shot up into the air. Looking down at Dau Tieng and the rubber plantation gave us a bird's eye view of our domain. We arrived about 10 minutes later at the grid coordinates and the pilot gently set the Huey down alongside the second bird (chopper).

SGT Spoores, along with the other NCO, directed our squads to set up a perimeter around the downed Cayuse. Both pilots were dead and still strapped in their seats. It was unclear if they were shot down but probably died from the crash. The chopper had ignited on fire by the collision with the ground and the airframe was burned to the ground. The smell of burnt flesh was something that I did not want to experience again. It was a smell that you just cannot describe, but know that it fights your senses and your state of mind. The pilots were still strapped into their seats and looked like charcoal briquettes starring out through the windshield. We hung around the site long enough for the graves registration detail to arrive by chopper and somehow pry the bodies of those two pilots out of the aircraft and into some black rubber body bags. That was a very somber moment to witness. We helped them load the bodies in their chopper and once they left the crash site, we waited at our PZ (pickup zone) for our ride back to basecamp. No one did any talking on the way back.

The 25th ID was made up of three brigades, the 1st, 2nd, and 3rd. Within each brigade there was an assortment of maneuvering battalions consisting of infantry and mechanized units. The mechanized (M) units were equipped with M113 APC's. The division had somewhere around 12 battalions assigned to it at any one time. These battalions were also called *organic units,* meaning that they were assigned directly to the 25th ID as part of their permanent makeup. The division strength stood at around 15,000 men, about a third of this number, were combat troops. As the 25th ID settled into this area, they assigned the 1st Bde to Tay Ninh, 2nd Bde to Cu Chi and the 3rd Bde to Dau Tieng where they would be headquartered. Each location was comprised of a basecamp large enough to house the troops and an airfield.

Remember as I describe this configuration, the 25th ID and 4th ID had each other's 3rd Bde under their control until that was rectified on 1 August 1967 when both divisions reflagged or swapped identities between these brigades. So, initially, it was 3rd Bde, 4th ID that was sent to Dau Tieng under the control of the 25th ID.

III Corp is a large area and required more than one division to fight the enemy. First to arrive in the area immediately near Saigon was the 1st ID. The following year the 25th ID began to arrive, followed by the 3rd Brigade of the 4th ID. The 25th ID was assigned operational control of the outlying provinces of Tay Ninh, Binh Long, Binh Duong and Hau Nghia. There were only two geographical points of note in this entire area, Black Virgin Mountain, called Nui Ba Den, 3,000 ft. high and home to our communications network, just north of Tay Ninh. Near Dau Tieng lies Pine Ridge, a part of the only mountain range in our AO. All the rest of the ground is flat, comprised of jungle, heavy brush, or rice paddies. In the middle of all of this runs the Vam Co Dong and Saigon Rivers. To the east of Dau Tieng is the Michelin Rubber Plantation which stretches east and north to the far end of the Razorback Mountains Ridge and toward An Loc and Loc Ninh, covering 50 square miles of trees and cover for the VC. There are other plantations to the west, Ben Cui and in the south; there is more around Cu Chi, Filhol Rubber Plantation and others.

Vast stretches of this area was simply uninhabited territory, especially all the lands west and to the north of Tay Ninh. As we moved over to Dau Tieng, the same results occurred. All the lands to its north were void of any occupants. Below the Saigon River, you could see very small hamlets, mostly occupied with numbers below 100 people.

Many areas of South Vietnam gained fame because of the fighting that took place or battles, large firefights, etc. In III Corp area, we had the Fishhook, Parrots Beak, and Angel Wings that define the shapes of the border between Cambodia and Vietnam. Within it, this tract of land, there were numerous geographical named areas: the Michelin Rubber Plantation, the Straight Edge woods below Tay Ninh, Black Virgin Mountain, also called Nui Bau Den, where the VC maintained some control over the lower part of the mountain for the entire year, the Trapezoid, Iron Triangle, Boi Loi Woods, Mushroom, Hobo Woods, Filhol Rubber Plantation, War Zone C and D, all major areas of constant clashes with the enemy and would remain so during the entire war.

5. MICHELIN RUBBER PLANTATION

We were located far away from any urban centers in the country. The two largest villages, the farthest northwest of Saigon, are Tay Ninh and Dau Tieng. Both of those villages hold hundreds of native locals whose main occupation is either farming, with some hope of making a few *Dong* (Vietnamese paper currency) at the local market, working for himself in some type of trade or as a field worker in the rubber plantations. To a lesser extent are the business shop owners in the village selling wares, food, and basic necessities of life. Away from the sounds of battle, the countryside is beautiful, with contrasting shades of greens set against the bright azure blue of the sky. As the sun traverses across the sky, large bold cumulus clouds form, dotting the horizon in various shades of white and gray. The day starts out with clear skies, but by early afternoon, the cloud cover builds and soon a heavy rain falls upon the landscape, basking the foliage in a refreshing cool covering of wet mist and rain that tempers the heat of the day and brings a different odor of freshness to the senses.

It was an odd and surreal scene to see farmers working the rice fields with their water buffalo and the children right alongside. Seeing a five or six-year old boy leading a 1,000 pound animal around without fear, yet for us, we were always willing to give these large beasts a lot of respect because we simply did not get along with them. Their nostrils flared and their tails whipped to and fro whenever we passed near them. It was always the prudent thing to do, to keep a wide berth when passing by one. I guess they just didn't like the smell of us or something, because we sure agitated them when we passed by. Momma-san would be busy in her mud and grass hooch, tending to the cooking fire and sweeping her mud floor of rice hulls and other debris that was tracked into her home. It was a simple life they lived, yet they were fearful of us and even more fearful of what the VC may do to them if they were caught giving us any kind of assistance. Either way, it appeared that they were in a lose-lose scenario.

It was mostly female workers that wandered throughout the vast forest of trees that made up the former French plantations where rubber trees were tended to. Each tree had a spiral cut in the bark and must be examined and occasionally recut to ensure a consistent flow of sap. The laborers emptied the small pots which hung from the base of the tree on a hook into larger pots which were then hauled away to be processed into raw latex. In the midst of these plantations resided small hamlets with names like AP 6, AP 13, AP 14, and none of them were happy when we showed up on patrol. They were strong supporters of the VC elements that passed through this section of Vietnam on their way toward the capital city of Saigon.

It became very clear to us replacements that there was something starkly wrong with how we were viewing this country. We could not figure out what the locals really wanted out of this war and exactly whose side they were on. I had only been there for three weeks and already the news we heard from home was upsetting. There were protests against the war all across the country, and we were wondering, what were we here for if our own country did not support the pain and suffering we are going through? In no time at all, rumors were floating around that we were entering peace negotiations with the

North Vietnamese government. We quickly began to think, "Hey, maybe we won't be here too long after all," but that was a falsehood and we regained our focus.

After the company returned to Dau Tieng, on April 9th, we were in rough shape having a number of wounded that needed some time to heal and were put on light duty, me being among them.

Two medics wandered by my hooch, opened the screen door and came in. Barry Olson and Jamie Ceballos introduced themselves. Barry was from Seattle, Washington and Jamie was from Southern California. Barry says, "You new guys should shed your underwear. It's going to cause chaffing between your legs with the upcoming Monsoon that will last for months and I don't want to be treating you for jungle rot. Make sure you keep your feet dry as much as possible and use that foot powder if you need to." I replied, "OK Doc, anything else?" Jamie piped in, "Yeh, don't forget to take those quinine pills either. No point in getting malaria here." We chit chat a bit more, then Barry said they need to go so they could make their rounds at the rest of the hooches.

Back on April 9th I write home to Mom:

Dear Mom and Dad:

Yesterday we had our first rain and it was a welcome sight. At least now we can breathe clean air and there isn't any dust blowing around. I received a pile of letters yesterday from Patty, you, Pat and Ruth, Kathy, and from those girls I met who live in Oakland (remember?). Anyway... Don't tell Patty or I'd be a dead duck.

With the rainy season upon us for the next six months or so, Charlie's activities should practically stop. He hates the rain more than we do. We're still humping the bush up in the Cambodia area. We have been clearing out an area where Charlie thought he could call home. Resistance is light and our morale high. That's the way it should be.

The PX here in Dau Tieng just doesn't have a damn thing in it. No cameras or radios although we should be resupplied in the next few weeks. I wonder how much these people appreciate what we're trying to do for them. They seem totally ignorant of that fact. Maybe this war will end soon and we can get going while the going's good. Ruth says it looks like they may have sold their Prince Avenue house since they took a $500 deposit on it the other day.

Routine here remains the same, up at six, out beating the bush by 9, till up to 4pm then digging our gun positions and catching chow in between until nine at night. Then we pull guard duty until 6am again. The days go by so fast all I can do is keep track of the number of the month. I just don't know what day of the week it is anymore. There's no difference between Sunday and Wednesday. The chaplain visits us when he can, regardless of the day. Must close take good care, love to you both.

Love your son, Arnold.

First rain of the wet season and somehow it was refreshing to smell the air dampened by the precipitation. While the rest of the battalion took up searching the rubber plantation to our east and north, we were relegated to providing security to the 65th Battalion Engineers who were making repairs to the major supply route in the area on Hwy 239 to the east of basecamp.

When patrolling the Michelin rubber plantation, you really had to be on your toes. The rubber trees come in all sizes. I don't know what the life expectancy is of a rubber tree but it must have one. Given the fact that each tree has a groove carved in its bark in order to collect the tree sap which is turned into rubber must cause a strain on the life of the tree. At any rate, the growth of brush and grasses even below the canopy of the trees inhibits your ability to see to relatively close distances of 50 meters or less. It was hot and humid and everyone was sweating profusely. Because I was still getting adjusted to the climate, I drank a lot of water. I also had a towel wrapped around my neck which I kept wet to try to keep my body cool. The sweat poured off my forehead at times because of the humidity. My shirt was saturated with sweat and my fatigue pants stuck to my thighs. After a week of this, the fatigues hung limp from my frame, unable to absorb any more sweat. The added tension of being on patrol and waiting for something to happen just generated more sweat.

On the morning of April 17th, we got into a small firefight with Charlie. We exchanged fire with each other. The *crack, crack, crack* heard from the occasional sniper fire did not allow us to pinpoint his or their location so we didn't know their strength. The enemy slipped back into cover, chuckling to himself that he had bested the American Imperialists. After several hours, the unit exited the canopy of trees and arrived on the outskirts of a village, one of many located in this vast forest of trees in the Michelin Plantation. We were spotted by some VC who fired at us. The company held its position as our commander tried to assess the threat. We began to take more fire from several points to our front. The request was made for gunship support from the 25th Aviation Battalion, their nickname *Diamondhead's*. Our C.O., 1LT Jay Hickey radioed the request to battalion who relayed the message on to Cu Chi. Battalion gave PFC Hiram Marziano, Hickey's RTO the Huey gunships, or LFT's "push". LFT stands for Light Fire Team. They were nearby and would arrive in 10 minutes. It would be easier to see what was going on from the air if we could force the VC to move.

Two Huey gunships arrived on station and began circling overhead. First platoon got a radio call from Hickey. They were to make a flanking maneuver to the left and attempt to flush the VC to our right. 1LT Ford's 2nd and 1LT Chris Brown's 3rd platoons were to hold positions for the time. The area was fairly covered with vegetation so caution was to be used. The sun was bright overhead and visibility was good, but not enough to spot the enemy's location unless we could see muzzle flashes from their AK-47's.

The gunships overhead asked us to mark our positions with smoke. 2nd platoon, in the most forward position, popped smoke and we got an acknowledgment from the air as Diamondhead 23 radioed back.

"Diamondhead 23, I see red smoke, over."

"Roger Diamondhead 23, this is Charlie 6 X-ray, red smoke, over" Marziano confirmed on the horn.

Meanwhile, 1LT Terry Keehn's 1st platoon had managed to circle the tree line leading up to the outskirts of the village. With the gunships overhead, the VC made a foolish mistake and fired their Chicom machine gun at one of the choppers. The tracers from the Chicom pinpointed the position of the VC and the second gunship, trailing the first, took aim and fired several 2.75 inch rockets at that location.

In the meantime, we continued to hold our position. There was no need to rush headfirst into trouble. The attack by the gunship had flushed the VC from their position and as we watched, they began to take turns making several steep dives toward the lush vegetation and palm trees concealing the enemy. More rockets were launched and struck the ground with a resounding explosion. This was followed by several bursts from a minigun. Within minutes the area was quiet. We didn't detect any movement, but continued to wait and see. Fifteen minutes went by and we got the word to advance. With the lead platoon approaching the village from the east and 1st platoon entering the village from the north, the peasants scattered as quickly as they could. We found two VC dead, hit by rocket fire. There was no sign of the rest of their unit. The C.O. released Diamondhead's gunships and gave them a "well done" for their efforts. The company quickly searched the village and found only a few old men and women in some huts. After we had taken a break, we left the area and continued our sweep back through the rubber trees and back to base for the night.

SGT Richard Thompson, 1st Plt told his story of this firefight after arriving in-country and recently joining the company. "When the firefight started, I hit the ground," Thompson said. "There were red and green tracers flying all over the place. I grabbed some grenades, and let them fly in the direction of the enemy fire."

Thompson remembers being spread out flat on the ground during this encounter. He continued his story, "After a brief few moments, Kjell Solberg showed up behind me and kicked my boots. I turned around to see Solberg standing over me, ignoring the ground fire that was going on, and said to me, 'Hey, why don't you join the rest of the company and get back where we are so the rest of us can join in on the shooting', then proceeded to walk back to where the company was." Thompson said he didn't realize that when he hit the ground nobody else did and he was all alone.

As I sat on my cot in our barracks tent that evening, I recounted the events of the day and the brief exchange of gunfire we experienced. I reflected back to April 4th, thinking of Schultz and whether I had done enough that day. I had my doubts about the judgment call I made in leaving him there, exposed to enemy fire lying on that road. I told myself that he was gone the moment he hit the ground, then reminded myself that I checked for a pulse and some sign of life. I had little time in doing this before I got hit. As I continued to wrestle with my memories, Mike Rinkle snapped me out of my thoughts. He was a tall slender kid from the Midwest with a good attitude, and infectious smile.

"Going to the chow halls to get something to eat?" he asked.

"Yeh, sounds like a good idea, let's go," I replied.

As we headed out the door and turned to the right we worked our way across the dirt road behind our tent hooch and through the rest of the company area that sat under the rubber tree canopy in our section of the basecamp. We slowly made our way to the mess hall.

"What did you think of Hickey calling in the gunships today?" I asked him.

"He seemed to really like to kick Charlie's ass with the hardware," Rinkle said, then continued on, "It's OK by me and better than us having to wade into those bushes getting someone shot."

"I agree," was my reply as I stepped through the door of the mess hall. "I hope we're not having shit on a shingle again. My guts still don't function very well."

"Nope, it looks like meatloaf and mashed potatoes, so you're safe for this meal," Rinkle said with a laugh, as we grabbed a tray and got into the serving line. The mess hall was bustling with activity. I was glad I didn't have mess hall duty anymore since leaving the States.

That same day, April 17th, Delta Company, led by CPT Edwin Bethea was caught in a crossfire in what was later called the Horseshoe Ambush. They suffered six KIA's and a dozen wounded. Their Medic, SP4 Ken Blakely was awarded the Silver Star for his bravery in moving through the intense fire giving aid to the wounded. Ken was a conscientious objector, which is why he was a medic, but had no problem learning and knowing how to fire a weapon. He was not afraid to defend himself if needed. He would perform more acts of bravery before his tour ended.

Tony Furrh remembers that day, "I saved one G.I. and pulled another to safety, but sadly the second did not survive. I had just placed them on flank when we made a quick boonie pit stop. I reacted to the sudden burst of fire from that direction by throwing off my radio and bolting into the triple canopy where I had last seen Lovick and Medina post out to flank. After an ensuing battle of sorts between us and snipers, I was able to save Lovick and at least bring Medina back for his family. We Medevac'd them out, and marched out as quickly as possible so that artillery and tanks could clear the area. We set up a few clicks away with a Mech unit, probably the 2/22nd. We were overrun that night. After events of the day, all I remember of the laager evening battle was that I was certain that we were not gonna make it out of this one. Guess the "*Man*" was looking over us that night. I think this is where on the casualty report for that day it shows Medina and, I think four other guys reported as KIA from the same area. I have discussed this with members of my company but most were pulled back to work on chopping out a Medevac site and were not exactly as involved as we were. 2nd platoon was on point that day and we had our hands full for a while and a man died and one was wounded."

Chester Bullard recounts with amazing clarity the battle of the Horse Shoe ambush and knew SGT Mouton personally, who was killed in this ambush..."When they brought him out from where he had gone down, they paused by my position to let me see him one last time since they knew we were sort of friends...he had a pen in the pen slot of his upper jungle fatigue shirt pocket, and I said, 'Well, let me take his pen so I'll have something to remember him by.' I pulled it out and found that it had been shot in two, only the top part or half of the pen came out. I kept it for many years but it must've gotten misplaced or lost as the years went by."

Bullard lost his hearing that day due to the fact that they called in an air strike and the Phantom jet came in so low he said he could almost read the writing on the bottom of it, and it dropped a 500 pound bomb that at once lifted him up off the ground and slammed him back down it was so close. After this episode he was totally deaf for a good while…of course they (command) wouldn't let you out of the field for anything, so it was some time before they actually believed him that he couldn't hear, like…going out on ambush at night and not getting the word that "we're going back in" and being almost left out there. They finally sent him to Cu Chi to have his ears checked and at that point the doctor there found he had ruptured eardrums. They put him on profile (removed from the field) for his remaining time which was only maybe two months or so. Prior to being removed from the field and before the Horseshoe Ambush, Bullard received an ARCOM with "V" for going down into a tunnel/bunker where he came face to face with two VC, one with an AK, another with a pistol that looked like a G.I. .45 but was perhaps Russian made…luckily he opened up with his M16 first. When they got the bodies out, they found that the grenade that had been thrown into the hole prior to going down into it, that a piece of shrapnel had dented the AK somehow and the one VC couldn't get it to fire. "It was jammed somehow," he said, "and the other one's pistol was out of ammo." I said, "My friend, somebody upstairs was looking out for you, that's all I can say."

It was April 19th, no one will say for sure, but scuttlebutt had it that our Battalion commander, LTC Charles Bauer had been relieved. Maybe his assignment was just temporary? LTC Donald Green was taking his place. He's tall, about 6'2", with a very slender frame. This was Green's second tour. He was here in 1964 as a Captain, serving as an advisor to an ARVN infantry battalion and task force. He speaks Vietnamese. With this background, it looked like we had a leader with some combat knowledge and experience. Prior to taking over the Battalion, Green served as our BN X.O., having arrived on March 3rd and was under LTC R. Dean Tice, until the 18th, when Bauer relieved Tice, then served under Bauer before relieving him.

In the morning, I was told to report to the First Sergeant, so after chow, I headed over to see him. SFC Harold Johnson told me he was sending me to be trained on the M2A1-7 Flamethrower. This unit was the same one used in WWII. It weighed 68 lbs filled and trying to maneuver wearing it was very difficult. It had a six cartridge ignitor that you triggered to light the jellied gas. Contrary to what was seen on TV, hitting the tanks with bullets did not explode the tanks, at least, that is what the instructor told us. As it was, I only carried it for one operation in the field and was glad to see the last of that weapon.

The Michelin Rubber Plantation was a good example of using dogs and their eyes, ears, and noses to help locate the enemy. In 1965, both the Marines and the Army initiated scout/patrol dog programs. The Army deployed IPSD teams in June 1966 at Tan Son Nhut AFB. In 1968, ten more teams were added and the last teams arrived in January of 1969. There were two IPSD's (Infantry Platoon Scout Dog) that we worked with, the 38th located in Cu Chi and the 44th located in Dau Tieng. Two other IPSD's operated within the 25th ID, the 40th in the Central Highlands at Pleiku then became part of the 4th ID and the 46th in Tay Ninh which joined the 25th ID after serving the 11th Armored Calvary Regiment sometime in 1968 following the Tet Offensive.

Trained to use their keen senses of smell, sight, and hearing, the dogs provided an early-warning system for the infantrymen. They were able to sniff out guerrillas, booby-traps, punji pits, and other potential dangers. To achieve the close rapport necessary between handler and dog, they were paired off at the start of the 13-week training period and were seldom far apart. They learned to understand each other's moods and actions, for in combat conditions, each depended upon the other for survival. Only if the handler was too badly wounded to continue, would the dog be taken over by another man. Arriving in Vietnam, the dogs were given 20 days to adjust to the weather.

"About the worst discomfort our dogs faced over here was the heat. They have two coats of fur, and need the adjustment period to shed their outer coat," explained 2LT Ian Jones of Houston, Texas, an IPSD platoon leader.

Where scout dogs had been used in Vietnam, they had proven so effective that the Viet Cong had standing orders to shoot the dog before engaging in any fight with the unit. The dogs even had their own equivalent of C-rations. While in the field, they were fed dog burgers, which are relatively compact to carry. In basecamp, they ate a mixture of dry meal and horsemeat. KP was pulled by the handlers, who fed, groomed, and cleaned up after the dogs. Grooming alone took up two hours a day, but it was all a labor of love. The men of the scout dog platoon knew the worth of their four-legged friends.

A request for a dog handler and scout dog was almost a daily request by Flame 6 (battalion call sign for the commanding officer) when we were stationed in Dau Tieng and working in the rubber plantations to the east and west of Camp Rainier. The underbrush which grew in some areas of the plantations was ideal cover for enemy movements and excellent cover to hide in from aerial observation. The scout team would always be up front of the column on point, working the area ahead looking for those telltale signs or scents of the enemy. One of the dogs we worked together with was Ringo, a big German Sheppard. The Sheppards were the primary dog used for this kind of work, using both their eyes and nose to spot danger. There were other dog units, known as Trackers. They were called IPCT platoons or Infantry Platoon Combat Trackers. The 66th IPCT was attached to the 25th ID. They used Labradors for their keen noses, to pursue and hunt down the enemy via blood trails.

During the Vietnam War 1960-'75, about 4,000 American war dogs were employed in various capacities, of these a few died early on in the war from food contamination; the Vietnam sub-tropical climate killed several hundred more. According to the Army Veterinary Corps, 109 war dogs died from heatstroke in 1969 alone; and from June 1970 thru to December 1972, 371 dogs were euthanized as being non-effective in combat, and another 148 died from various causes; during the entire war 288 were officially listed as killed in action...along with 285 handlers.

More than 9,000 Army, Navy, Marine, and Air Force handlers served in Vietnam during America's involvement. Were the dogs of Vietnam effective? Our military experts and "armchair Generals" will probably be debating that question for the next hundred years. But any Vietnam combat veteran that happened to be part of a patrol that was saved from a VC ambush because of a scout dog's alert or prevented from walking into a mine field, will tell you, the answer is definitely yes! The Viet Cong thought so too...they placed bounties on both, the American handlers and their war dogs! Estimates vary, but some state that the dogs may have been responsible for the saving of at least 10,000 lives in Vietnam.

Note: A few quotes in this chapter from K-9 History, the Dogs of War and another story reference is http://vdha.us/ Vietnam Dog Handler Association.

6. RADIO TELEPHONE TRAFFIC AND LANGUAGE

For the 25th ID and other line units, air mobility was an important factor in catching the enemy off guard and due to terrain conditions sometimes necessary just for the sake of expediency. Our Brigade made many helicopter assaults in Vietnam. We called them CA's—short for combat assault and eagle flights—yet many veterans don't know how complicated an operation they were.

This chapter is written to illustrate all that takes place in the planning and execution of a mission from the viewpoint of the company commander of Charlie Company (Charlie 6 is his radio call sign) who is receiving orders from the Battalion Commander (Flame 6) or his S2 (intelligence) and S3 (operations) HHC staff. A typical CA might start with a morning radio call from the boss, a bird colonel who is in charge of the brigade (COL -full colonel) or a light colonel (LTC—lieutenant colonel), the commander of the 2/12th Bn. The Platoon Leader or the RTO was responsible to carry the Signal Operating Instructions (SOI) which contained the frequencies and call signs of all units in the area and day codes that may be needed to conduct the operation. SOI's enabled the RTO to encrypt sent messages and receive coded messages, then decrypt them. The SOI provided the organization of stations into nets, assigned call signs, designated net control stations (NCS), and assigned frequencies. It also provided information on changes to alternate frequencies and on authentication codes. In addition, the security procedures that must be used by radio operators in the command are included in the SOI supplemental instructions.

Note: Initially, when I assumed duties as an RTO (June to August 1968), many of these added measures were not in place. We did not encrypt our map coordinates nor some of the air traffic call signs until mid to late 1968. We pretty much used open language to communicate.

The use of the radio changed over time to prevent the enemy from intercepting messages and using them against us. We started to code some of our messages using three letter phrases. We also resorted to challenging someone receiving our transmissions by stating "authenticate" and the receiver would have to transmit back a password, or code word to verify they were legitimate. Grid coordinates were no longer transmitted as XT924211, but used something called SHACKLES. These were a series of 10 letter codes for 0 through 9 to translate map coordinates. Each number was represented by a letter and the sender and receiver RTO would each have a code book that was changed weekly to be used to interpret the message. So, let's follow a series of transmissions, before code books were used, to demonstrate how field forces were moved from one spot to another. It might sound something like this:

"Charlie Six, this is Flame Six, over."

RTO (Radio Operator) would acknowledge the call by saying "Flame 6, this is Charlie 6 X-ray (X-ray is the RTO) — stand by one," then give the "horn" (radio handset) to the C.O. "Flame 6 — this is Charlie 6. Over."

"Charlie 6. Fullback 6 (call sign for 2/22nd Mech BN CMDR) had some minor activity last night. We're going to drop you in west of his AO and see what you can stir up. The Sierra 3 (S3 is the battalion operations officer) is trying to get some birds (Hueys or slicks) to move you. We'll give you an ETA when we can. We've found a Papa Zulu (pick up zone) for you — have your 6 X-ray stand by with his SOI. Over."

"Flame 6 — Roger. When do you want us to move to the Papa Zulu? Over."

"Charlie 6 — Begin your movement when you're ready. Over."

"Flame 6 — Roger — ready with the SOI. Standing by. Over."

This bit of military radio-speak was a fragmentary order — or, as we called it, a frag order — meant to give us the "heads up" to prepare for a helicopter combat assault. The battalion commander told us that Fullback 6 had a minor exchange of gunfire during the night, indicating the enemy was in one of his Company's area. The boss wanted us to move to a position west of 2/22nd in an attempt to locate the enemy. To do that, we would need helicopters. The commander's staff officer (a captain) in charge of coordinating combat assaults was the S3 Air, and the colonel had him calling division headquarters trying to schedule some helicopters for us. While he was busy doing this, we would *hump* or walk through the jungle towards a clearing Flame 6 had spotted from the air. This clearing was the *Papa Zulu*, — radio speak for the letters P and Z, indicating a pickup zone. It was the place helicopters would pick us up to ferry us to the new place — known as the *Lima Zulu*, or Landing Zone, west of the 2/22nd. He was going to give us the map coordinates of the LZ in code so the enemy wouldn't await us in the clearing. That's why he waited for the RTO to have his SOI ready. This little code book was worn in a small waterproof pouch and was the only classified material carried in the field.

RTO's were sharp people. In a lot of ways, they functioned as an alternate commander. As the company commander, there were two RTO's — one for the battalion network connecting the TOC or tactical operations center, the battalion commander, and some of his staff, and one for the company network (company commanders, the platoon leaders and sergeants, the First Sergeant, etc.). These guys were not specially trained to be RTO's—they were just smart and dedicated and learned their craft on the job. Besides being smart, they had to be strong. Not only did they carry all the gear the rest of the soldiers carried, but also carried the 30 plus pound radio—and even spare batteries. A close bond existed between the commanders and their RTO's.

Having overheard the conversation with the battalion commander, the RTO with the company radio would have anticipated the need to see the platoon leaders to give them a *frag order*. In a few moments, the three platoon leaders, their platoon sergeants, the artillery forward observer who also had his own RTO to talk on the artillery net to direct fire control, would be standing around awaiting orders. It might sound like this.

"Fullback had some contact last night and Flame 6 wants to insert us west of their AO. From looking at the map, that should put us near AP 2, north about two clicks. Our PZ is at map grid XT554520. We don't have the birds yet, but the S3 Air will let us know when they have an ETA. Let's saddle up in three zero. We'll move in a column of twos with flankers out, with 36 (third platoon) in front, then HHC, 16 and 26. 36—when you get to the PZ, hold up short and let me know. When we get the birds, we'll have 36 secure, with 16 going first, then 26, with 36 on the last lift. Focus 28 (F.O.)—go out with 16; I'll go out with 26. Questions?"

Usually, there were no questions because a frag order was just that—it gave only a fragment of the information needed to carry out the mission. These men knew they would get more information. In the meantime, each had work to do. The first was to move from our current location to the PZ Flame 6 had located for us. The company had 30 minutes to start moving, and there was much to be done.

The Platoon Leaders were usually young lieutenants with little military experience except for OCS (officer candidate school), maybe Ranger school and six weeks of jungle training in Panama before arriving in Vietnam. Each commanded a platoon of about 25 to 35 men. They received strong support from their more experienced Platoon Sergeants. A platoon had four squads, each led by a squad leader, squad 1 to 3 being a rifle squad and the 4th, a heavy weapons squad consisting of two M60 machine gun crews. The lieutenants would go back to their platoons, gather the squad leaders around and give them a frag order. He would tell them the same basic information I gave them, but would also tell them the position of each of the squads during the hump to the PZ, give a compass azimuth (direction) to the squad leaders, and other information. The platoon sergeant would work with the squad leaders and assign soldiers to the helicopters. As each Huey helicopter, nicknamed a slick because it lacked armament, could carry six fully equipped soldiers, care had to be taken to divide the troops into proper sized groups without breaking up the squads in such a way that they wouldn't be able to work together as soon as they were on the ground. There was also the problem of not overloading a helicopter. Air density was a factor in Vietnam because at times the Huey struggled to gain lift and get off the ground with its load capacity. We experienced times when we had to offload weight to get a "ship" airborne and at critical times.

The F.O. was a lieutenant with the responsibility of directing the firing of artillery howitzers located at various fire bases around us. Along with his recon sergeant and RTO, he would insure we had defensive fires to help us should we be attacked at the PZ or anywhere else, and he would coordinate with the captain assigned as the artillery liaison officer at battalion level to be sure he knew what fires would be laid down at the LZ just before we landed in the new area. The artillery folks were attached or assigned to any unit requesting their support, take orders from that unit but were not part of the organizational unit, Flame 6 in this case — they were members of Battery A, 2nd Battalion, 77th Field Artillery. The F.O.'s call sign was Focus 28, and he stayed with the company commander. The Recon Sergeant could operate independently when the company was split using the call sign Focus 28 Delta.

Platoon Sergeants would assign the individual soldiers to helicopters. We would have six slicks move us, and they would do it in shuttle fashion. Each shuttle was called a *lift*, therefore first platoon would go on the first lift, second platoon on the second lift, and third platoon on the last lift or the platoon sequence could be altered based on other needs, i.e., experience in leadership or a specific platoon assigned to take point (lead the company in executing the mission). This scenario or others, all depended on how many air assets (helicopters) we were granted for an operation. The more Hueys the less lifts we had to account for. A complicating factor was fitting in the company headquarters, artillery personnel, and sometimes, but rarely, part of the mortar squad.

After a couple of hours of walking, the 3rd platoon leader (36) has his platoon at a big open field. He stopped his troops and advised the C.O. they were at the PZ and the C.O. usually confirmed his navigation, and the company was told to stop and put out some security. After a radio call to battalion, we were told we had helicopters scheduled with an estimated time of arrival (ETA) within one hour. The C.O ordered the unit to take cover as he didn't want anyone out in the open field for fear of letting the enemy know we were there and would soon be picked up by helicopters. The big aluminum birds were easy targets just before they touched down to pick up the troops, so we waited for the last possible moment to enter the open PZ. In the meantime, Flame 6, his S3, and the artillery liaison officer were up in the air in the commander's C&C, command and control helicopter.

They would oversee the pickup and CA from 5,000 feet up and coordinate the activities of Fullback 6, the artillery, the troop carrying slicks, and the helicopter gunships. Soon, a call from Flame 6:

"Charlie 6 — this is Flame 6. Over."

"Flame 6 — this is Charlie 6. Over."

"Charlie 6 — Standby to copy the location of your Lima Zulu. Over."

The battalion commander then transmitted a coded message with the map coordinates of our LZ along with the direction he wanted us to travel once we were on the ground. That gave the C.O. a chance to call for the platoon leaders of Charlie 6 and give them the rest of the information they did not get during the first frag order.

"Charlie 6 — Your birds are in the air inbound your location. ETA one five minutes. Over."

"Charlie 6 — Roger. Moving onto the PZ. Over."

"Flame 6 — Roger. Out."

With that, the C.O. got on the company radio network, telling the platoon leaders and others to move out onto the PZ. While it might appear to be a bit chaotic, each group moved according to the plans laid out in the frag order. The third platoon — the one that had been leading the hump — would split. One squad would walk through the wood line at the edge of the PZ, and the other two in the other direction, also just inside the wood line. Their job was to build a big circle. Once they had done that, they would move into the field, but only a short distance in from the wood line. By looking outward, they would provide protection to the rest of us while we boarded the first two lifts of helicopters. In most cases, the artillery recon sergeant, Focus 28 Delta, would also be with the last platoon to be sure someone could direct artillery fire should the PZ get hot.

Third platoon would just about have the ring closed when first platoon would enter the open field. They would split into six groups—one for each of the six slicks—as they would be the first group to be picked up and taken to the LZ. The 1st Platoon Sergeant would have also made room for the F.O. and his RTO. His presence was to insure we had someone on the ground able to adjust artillery fire should the LZ be hot.

The second platoon would also enter the LZ, but would spread out a bit more and place themselves between the six groups of 1st platoon and the protective circle of 3rd platoon. The Platoon Sergeant would split up the C.O. and PL and their RTO's between two birds in the second lift, and find a place for the company senior medic as well. While all this was going on, the helicopters were on the way. We were often supported by the 187th AHC *Crusaders*. The commander of the troop-carrying slicks would have changed the frequency of his radio and called us on our company frequency (called a push or freak). It might sound something like this.

"Charlie 6 X-ray—This is Crusader Yellow One on your push. We're inbound your location. Over."

"Crusader Yellow One—Charlie 6 X-ray. Roger. What's your ETA? Over."

"Charlie 6 X-ray—We're about zero five out—go ahead and pop smoke, over."

"Crusader Yellow One—standby. . . . smoke out, over."

At this, the RTO, or someone else, had thrown out a smoke grenade. It produced a lot of billowing colored smoke. The purpose was twofold: it would be a method of identifying us and it would tell the helicopter pilots our position and the wind direction. To prevent the enemy from fooling the pilots into landing in the wrong place, the ground troops always threw the smoke, and then asked the pilots to identify the color.

"Roger—Charlie 6—I've got purple, over," replied the lead Huey pilot.

Charlie 6 X-ray responded, "Affirmative on the purple smoke, Crusader. Charlie 6 X-ray, out."

About that time, six slicks would appear overhead. They would usually be flying in two columns of three slicks each, one slightly behind the other. Formations would vary according to the situation and would be dictated by the flight commander. The six groups of 1st platoon would be arranged in the same formation, with each group separated by the same distance as the helicopters. The troop formations would be far enough apart for the six slicks to land between them. Loading of the slicks would be done from the troop side of the chopper. This was done whether to on load or offload, so we could get into a defensive position with everyone close to each other and able to effect fire in a circle keeping the choppers to our backs. The slicks would be escorted by helicopter gunships if they expected enemy action or a hot LZ or PZ. These fearsome critters were capable of bringing pure hell down upon the enemy. They were armed with 2.75 inch rockets, some with automatic 40mm grenade launchers, and a minigun. This special machine gun was a modern adaptation of the multi-barreled Gatling gun. They would not use those guns at the PZ for fear of hitting friendly troops, but they were ready if we needed their help. As the slicks made their last lazy-looking turn into their "final," one trooper in each group would stand up in the field with his M16 rifle over his head. His job was to provide a guide for the helicopter pilot picking up his group. Even though he might make motions with his arms in an effort to help the pilots, I always felt the pilots basically ignored them and just used them as a guidepost for landing. The birds would not be on the ground long—the troops would start running in a crouch for the open side doors before the skids even touched the ground. Being careful to avoid the tail rotor, they clambered on quickly. The helicopter crew chief and door gunner would tell the pilot when to "pull pitch", and off they'd go into the sky.

The doors were never closed, and nobody ever fastened seat belts. It was too loud to talk much, but the air was so nice and cool up above that jungle.

But while the lead platoon was in the air, and the rest of us sat on the PZ, other folks were busy. Up in the "C n C", the artillery officer was bringing steel rain down on the new Landing Zone.

His job was to saturate the rim around the open field with howitzer fire in an attempt to kill any enemy that might be lying in wait for the incoming helicopters. At a minimum, he would have six 105mm guns firing, and often had more. An artillery battery was stationed on each fire support base. A battery consisted of six howitzers. He would be firing traditional artillery that exploded on impact with the ground, but he would also call for some of the shells to explode above the ground, causing hot shrapnel to spray the area. The "LZ prep" would go on for 15 or 20 minutes, and the gun crews would be firing the guns just as fast as they could stuff them with fresh rounds. It was a hellish display of firepower. If you watched an LZ prep, you wondered how anything could possibly live through such an ordeal, but the enemy often did.

Of course the helicopters had to steer clear of this steel rain, so the assistant flight commander would be listening to the artillery radio network and would be aware of the trajectory of the shells. At this point, a CA called for perfect coordination. The idea of the LZ prep was to keep the enemy's head down. If there was too much of a time lapse between the cessation of the artillery fire and the landing of the slicks with their load of troops, the enemy would have time to get weapons in place and shoot at the helicopters and the troops. If the helicopters tried to land while the artillery was still being fired, they ran the risk of being hit by a shell. Timing had to be perfect. By listening to the artillery radio net, the helicopters knew when the prep was about to end.

When that last round was fired, the message "tubes are cold" was sent out, and the helicopters moved in close. It was desirable that the slicks be near the LZ when the last round went off. Once the artillery fire was "lifted" the gunships would then fly in low with rockets, grenade launchers, and miniguns going. They would rake the wood line as the slicks approached. As they got near the ground, the M60 machine guns mounted on the slicks would also open up. The noise level approached the unbelievable. The troops were also itchy to get off that big flying target, so they would often be outside the helicopter, standing on the skids. As the slicks approached the field, they would "flare out" just before touching the ground. Normally, the skids never touched the ground—the troops jumped off and ran towards the tree line. The M60s on the slicks were now quiet because there were friendly troops on the ground, but now the grunts opened up with their own weapons.

The gunships now had to stop firing for fear of hitting the troops. The slicks were now gone. If there was no enemy fire, the platoon leader would radio and inform me the LZ is green. Of course, everyone else was listening in as well. The lieutenant's job was now to organize the men on the ground in a circle and secure the field for the rest of us to land. The quiet was deafening—and welcome.

Now the shuttle began—the birds went back to pick up the second lift, and then the third. If there was no enemy contact, we were all one happy family again. Within 30 minutes or so, we would be heading off that open field in the direction Flame Six wanted us to search. And that is how an airlift was conducted.

7. EAGLE FLIGHT

April 18th we caught an eagle flight from the 187th AHC, Assault Helicopter Company, the Crusaders out of Tay Ninh. They had a nickname for us, the *Electric Strawberries*, but they were the only ones who used it. We were airlifted out of Dau Tieng and joined up with the rest of the battalion about six clicks northeast of basecamp. That was on the northern border of the Michelin rubber plantation and we were at the foot of a string of mountains that divided this area known as the Razorback Mountains. On the northwest side of the mountains lies War Zone C and to the southeast lies the Michelin rubber plantation. We were located on the southeast slope of Pine Ridge.

The jungle was thick and dense and visibility was nonexistent. We came across a grove of bamboo and the size was indescribable. Each shoot of bamboo must have been four to five inches across. As we moved through the heavy underbrush in the jungle, I spotted a scorpion, black in color and the size of my hand. He was in a very defensive posture and everybody around decided not to mess with him. We moved through a high grove of elephant grass which is razor-sharp and we were careful not to lacerate our skin or arms. Any aggressive movement through this stuff would result in torn skin. The blades of grass are like small saw blades. We stopped for a break and found a spot to sit amongst the tall strands of bright green grass, watching as it gently bent with the light breeze being channeled off the slope of the mountain. The air was heavy and the humidity so high that we struggled to get our breath. Sweat was just pouring off me and my fatigues were so damp that they wouldn't dry out. After our break, we pushed on with our patrol, but found no evidence of the enemy. That afternoon we set up a laager site and settled in for the night, awaiting the possible arrival of the VC probing our defensive perimeter.

We conducted daylight AP's (ambush patrols) in the area for several days. During one of these operations, we had taken some gunfire from a hedgerow to our front. 1LT Jay Hickey, our C.O., ordered the company to get online and that we were going to assault the hedgerow. As the command was given, we all jumped up and started to march forward.

PFC Sidney Fowler, who arrived at the same time as many of us in March, was a rifleman from Georgia. For whatever reason, he jumped out in front of everyone and started to sprint across the formation about the same time as we started to fire into the wood line. PFC Barry Price, an M60 gunner started to light up his "60" when Fowler jumped into his line of fire. One round from the machine gun hit the back of Fowler's steel pot and he tumbled to the ground.

Holy Shit! Price thought he had killed the guy and quickly told his ammo bearer not to say anything. Not say anything? Hell, Price just shot the guy in front of the whole company. Well, Fowler was only grazed by the bullet. Once it hit the helmet, the bullet deflected downward, grazing Sidney in the process. They Medevac'd him back to Dau Tieng to get patched up. While he was there healing up, he got the notion that he should have all his teeth pulled since they weren't in good shape anyway. Apparently, some army regulation says no teeth, no frontline duty. After that, Sidney was reassigned to resupply support and served the rest of his tour in that capacity. Being assigned to resupply meant he was no longer out in the field in combat, but stayed back at our basecamp. The job involved bringing food, ammunition, ordnance for the mortars, the mail etc., out to where the field troops were. It seemed months, before I saw him again when he showed up with a resupply convoy out in the field riding in a Jeep. No one ever told him what really happened that day.

Each night we set up a new night laager and waited for the arrival of the night kit so we could set up our perimeter. The CH-47 Chinook arrived, assigned to the 242nd Assault Helicopter Support Company or ASHC for short; their nickname *Mule Skinners*. They dropped off a water trailer, left and returned with the night kit, rolls of concertina wire, Claymore mines, and a hot meal. The meals were stored in Mermite containers. These are large insulated thermos metal boxes with a lid that attaches with four snap hinges. Some were single storage compartments and others had dividers for multiple compartments. For a treat, we also got two pallets of refreshments, one of beer and the other soda. The beer could be Pabst Blue Ribbon, Schlitz, Budweiser, or maybe Hamm's, along with Coke, Pepsi, or 7-UP. The only problem is they were hot and who wants to drink hot beer or soda. Each man was allotted two beers and two sodas. This practice of sending beer and soda to the field ended as a battalion practice a few short months later, never to be done again.

It was time to get serious about setting up our night laager site after chow. We were sitting in a clearing surrounded by heavy tree cover, burning a few cigarettes and waiting for instructions from 1LT Vandervoort. Hickey alerted the platoon leaders to where their platoons would set up, then the word filtered down to our squad leaders what section of the perimeter third platoon would have responsibility for. Our field of fire to the edges of this area was about 50 meters. Too close for comfort but better than sitting somewhere where our line of sight was blurred by underbrush and growth for cover. It was very dark once the sun fell below the horizon. We hurried to get our defenses set up. This meant stringing out the coils of concertina wire around our perimeter. If we had enough, we would string two rows, hopefully slowing down the enemy if he should choose to assault our position during the night.

"Let's get hopping boys, get that wire strung out fast and set our Claymore mines up," Spoores said.

These are anti-personnel devices which will blast out a pattern of shrapnel in a 60 degree direction, effectively killing anyone within five meters to the front of the mine. They have a blasting cap and wire which is strung back on the ground to a foxhole where the wire is attached to a hand held device, a detonator, which when squeezed, generates an electrical charge setting off the blasting cap and the 24 oz. C4 plastic explosive inside the mine. This mine was invented by a guy named Norman MacLeod and named after a large Scottish medieval sword. It fires steel balls out about 100 meters in a 60 degree horizontal arc. The range for achieving a kill zone or causality radius is about 50 meters. They are a nasty but effective weapon.

"Krause, you and Price handle the Claymores on left side and Hill, you and Potts set up the right side. Make sure you string the charge wires back to each firing position and set up the detonators. The rest of you work the concertina wire into place."

Price and I pulled the Claymores out of their bandoleers which contain the mine, command detonator, and wire with the attached blasting cap and went to work. Each mine has the words "FRONT TOWARD THE ENEMY" embossed on the front of the mine, just to make certain they were pointed in the right direction.

If we had time, we would take some empty soda or beer cans and put rocks in them, then hang them on the wire. When the wire was bumped, the can would rattle, signaling that we had an intruder. When that happened, a call to the F.O. and to the supporting artillery battery would request some *illumination* rounds to be fired above our location. These aerial flares are suspended by a small parachute and slowly drift earthward, lighting up the countryside and exposing the enemy to small arms fire. The flares burn for 5-7 minutes. When a flare would detonate overhead, it would "pop" and the flare suspended from a small parachute would ignite, then you would hear a slight "PHEW, PHEw, Phew, phew, phew" as the cylinder separates and starts to gravitate to the earth.

Illumination rounds were mostly provided by 105mm and 155mm howitzers or four deuce mortars if we were within range. If air support was on station, a C-123 flare ship could also be dropping illumination if needed when units were under ground assault. Either way, each flare would burn for five minutes or so before extinguishing. Also available for illumination, were hand held flares that were aluminum tubes about 1 ½" x 12" in size. They had a removable cap which was removed and slid over the bottom of the tube which had a firing cap. The cap had a detonator pin inside and when the cap was struck by striking it with the hand, knee or hard surface, would set off the flare which was pointed toward the sky. These types of small flares would provide about a minute or two of illumination.

"No trip flares tonight," Spoores adds as he watches and offers instructions to his squad. "It's too late to be able to set them up without setting one off," as he finishes his thoughts. Trip flares were small in size, about 1 1/2" x 5" and designed to use a trip wire strung across trails or other avenues of approach. Once the wire was tugged on, the pin holding a spring-loaded pin detonator would release and strike the cap igniting the flare which would light up a large area for everyone to see. There was no hiding from sight once they went off. They put out about 35,000 candlepower and burned for a minute.

After we had set up the perimeter, it was time to dig a firing position or some sort of foxhole to protect ourselves from incoming mortar rounds which would invariably always show up sometime in the evening. The VC were very deft at finding out where we were, and they were accurate gunners with their 82mm mortars. The ground was hard and it was difficult to dig, but we tried to get underground as deep as we could. Our survival depended on this and that point was reinforced by our squad leader, repeatedly.

As soon as darkness set in, we set up security. Each squad in each platoon had someone watching the wood line. If we could, we sent out listening posts or LP's in several directions from the laager site. They were three men teams who found a spot out about 100-150 meters from our location. Their job was to be our eyes and ears to detect any massing or movement of enemy troops in the area. If so, they reported back by radio, then depending on their safety or exposure would stay or retreat back to the laager site. Tonight, we had no LP's out; it was too risky.

"Keep the chatter down, pass it along," whispered Spoores. "We don't want to help the VC find us out here, and stay alert when you have the *watch*."

Everyone was nodding off to sleep except for the new guys who were seeing and hearing all kinds of stuff. Their imaginations were on overdrive. Sweat was forming on their foreheads, and their armpits were dripping wet. Every sound detected was a VC. They got little sleep, and that included me.

Off in the distance, there was a faint *bloop, bloop, bloop, bloop* heard and SGT Spoores yelled out, "*INCOMING*" along with other veterans. Hearing those words shouted out meant we had 5-10 seconds to find a deep hole and get in it before the first mortar round hit our laager site. The explosion was loud, and grew louder as each successive round marched closer to our position. I was hunkered down with William Potts as we held onto our steel helmets.

The VC sent their mortar rounds in an arch, hoping to hit our area. This is called *walking* the rounds in. He would send a dozen or so rounds at us, then run for cover. If we could use counter measures, if we had our own 81mm mortar crews with us, they would quickly calculate the range and distance they think the enemy gunners were firing from and return fire, hoping to catch them in the act before they could retreat. It was a game of cat and mouse. We were good at it, but so was *Charlie*. If we got more than a dozen or so rounds, it could mean they were attempting to soften up our positions and hit us with a ground attack, but the firing ceased and soon it became eerily quiet. A quick check and for tonight, no one was injured. The incoming rounds hit on the outside of the wire. The rest of the night remained solemn. A routine we would witness countless times over the coming weeks and months.

On April 21st at 0220 hrs., while pulling a company ambush patrol, we sprang the trap on 20-30 VC, opening up on them with small arms and automatic weapons fire and Claymore mines. In the process, six VC were killed and we took one POW. The next day, April 22nd, we moved further north near a small spot on the map called Xam Bau Dau. We had another run-in with Charlie and PFC John Taitague from Guam was killed. This was still heavily forested area that we were in. Contact was quickly broken off by the VC before we could retaliate. It was very easy for the enemy to melt back into hiding and disappear from sight.

Someone called in a Dustoff and our platoon leader was taken into Cu Chi for heat exhaustion. He never returned to the company. Later, we learned he was reassigned to 3/22nd. 1LT Chris Brown, a man not much older than us, from Texas, was flown out to the field to assume command of 3rd Platoon. After continuing to search this area, all companies were returned to Dau Tieng by truck on April 24th. During one of these trips, we had a reporter from the Tropic Lightning newspaper assigned to the battalion. He got a chance to snap a few pictures. One of those is of Price and Christianson posing like they were covering the unit with their M60 machine gun.

Back at basecamp on April 24th around 0130 hrs, we were put on alert status and grabbed our gear and headed to the airfield. A unit in an NL nearby was in contact with the VC, and we waited to be sent out to reinforce their night laager sight. We sat by the airstrip, pondering what it might be like to make a night combat assault. As I propped myself up on the ground next to the airstrip, I reminded myself that I had been in country just over a month now. Already I had been in numerous firefights and had just about experienced enough to call myself a veteran. Thinking back on some of my *firsts* gives me time to reflect on how I have handled each situation.

The first time I went out on a night ambush patrol pushed my imagination button to full on. The heart was pounding as the eyes and mind wanted to fabricate an enemy soldier out of every shadow, bush, or sound that was detected as the patrol slowly moved through the terrain on its way to the ambush checkpoint. Once the patrol had arrived, each man carrying a Claymore mine set it up so it would create a killing zone across the path or oxcart trail we had staked out. Each squad took its turn having someone on watch, looking for some unsuspecting person or persons to wander by our location.

Vietnam had a curfew, which was dusk, and anyone caught moving about after dark was considered to be the enemy and was fair game. It was hard to stay awake for two hours after being up all day and everyone fought the urge to close the eyes and fall to sleep, except the new guy, me and a few others. We had the mojo going with our imagination, and that alone scared the hell out of us. We awaited the prospects of triggering an ambush and the ensuing firefight. It would be a restless night.

Richard Polus, from Indiana and George Reitman were sent out to the field, a couple of replacements. The new guys were picked on first by their squad leaders. The tasks least liked on patrol was being on point or the first person in the formation on patrol or the flanker which was the side security each patrol used to have eyes and ears out on each side of a column. There were two flankers to the left and right of a column and they were usually about 25 meters away. It was their job to look for trouble from the sides while the point man was concentrating on the same at the front of the column. Either position was extra dangerous to one's health, and was usually the focal point of enemy fire, should that opportunity present itself. Most veteran squad leaders would put the new guy out on flank rather than expose a veteran, but the down side to that, was the new guy might not know what to do if fired on, or saw something suspicious and did not take the appropriate action. A bad choice was not good for him and certainly unhealthy for the safety of the unit he was out there to protect. A smart sergeant would split the difference and given the choice, would at times ask the veterans, or if needed, say who would take point and who would be flanker. He had to use his instincts and trust them to make the right decisions. I'm brought back to the present when at 0300 hrs we were notified to stand down and we returned to our company area and tried to get some sleep before we had to roll out of our sacks at 0600 hrs.

Each day was like a repeat of the previous one, but monotony was not part of anyone's order of the day. All it takes just once was to drop your guard and it could be all over for you. The rest of the month kept us scouring the rubber plantation to the east and repeating visits to the small villages that couldn't wait to see us so they could give us the cold shoulder. The S3 operations officer even assigned us as a security detail to protect some 65th Battalion Engineers who were working to replace a bridge on Hwy 239 for a few days. That was some easy duty that allowed us to catch up on some rest.

During the next week, we systematically performed cordon and search operations on the villages in the Don Diem Michelin Rubber Plantation. A cordon and search worked this way. We would slip into an area near a village, then deftly surround the village, blocking all exit avenues, all done in the very early hours before daylight. At dawn we would enter the village and question the occupants while checking identification papers and hoping to trap VC who entered the village at dusk. They could be visiting family or friends at the time and we would take them into custody or if they ran, it was at their own peril.

After two days of patrols out of Dau Tieng, on April 26th we combat assaulted three companies to the northeast and set up an NL. On the 28th we conducted a cordon and search operation using three companies at village AP 6 Chanh, but found little. That night our laager site received 25 rounds of 82mm from the local VC. This resulted in five U.S. WIA's of which we had to Medevac three back to Dau Tieng for treatment. We continued to conduct operations in this area for the next seven days before returning to basecamp.

The morning that we returned to Dau Tieng started out in routine fashion. Everybody knew what needed to be done when we had to pack up the night kit and made it ready to be picked up by the CH-47 Chinooks. That meant rolling up the concertina wire, packing up the Claymore mines, putting the PSP (perforated steel plate) and bundles and securing them in cargo nets. PSP was designed to interlock and used to build temporary runways. But, it was also good to be used as additional protection from mortars when put on the top of a firing bunker, and covering it with sandbags. We usually had access to this material when we built bases intended to be occupied for extended periods.

Along with all of this we had to pack up our ammunition and grenades, mortar rounds, and any other explosives we may have. As we were awaiting the arrival of our first pick up from a Chinook, the order was given to pop smoke. Unfortunately, the G.I. tossed a smoke grenade right on top of the cargo net containing all our ammunition. Smoke flares put out a lot of heat and have a flame and when you combine those with ammunition, good things don't happen. The ammunition and mortar rounds caught fire from the smoke grenade and started igniting. Everyone was diving for cover as bullets were whistling everywhere and the mortar rounds were exploding. We found ourselves under cover for several hours as the heat cooked off rounds of every variety and sent them in every direction. Fortunately from what I can remember no one was hurt, but there was a lot of explaining to be done. It was quite a sight to see.

There were a number of small villages that surrounded the countryside. The ones in the rubber plantation don't have names but are referred to as AP 6, AP 12, and AP 14, etc. AP is the designation for a small hamlet. We eventually seemed to visit them all. The local people took our presence with calm indifference. You had to realize that the people there had been under constant war since before WWII. After the World War, it was the French attempting to regain control of the country against Ho Chi Minh and his army, and then we stepped into the picture in the early 60's.

So, as I was describing the local peasants, they tolerated us and our bumbling attempts to maintain control. The truth of the matter was, we only controlled the ground we stood on, and where we had established basecamps or fire support bases. Even the ARVN's lived in compounds, and did not mix with the local population. I guess they either feared for their own welfare, or were facing the same dilemma we were in, not knowing who the enemy was. Fine you say, and true as long as we were confronting the NVA. They were nice enough to wear actual military uniforms, including pith helmets. But Charlie, the black pajama VC, was another story. He looked like any other farmer who casually went to market or was busy tending to his rice paddies or family. We didn't want to trust anyone nor turn our backs on these people.

Aside from this situation, the kids were great. They loved the G.I. and were happy to hang around us when given the chance. Everywhere we went, there were kids trying to sell us iced sodas, beers, and local food as they trailed us on bicycles and afoot. A soda usually cost 50 cents and were always hot. You could get a block of ice to cool down the drink for another 50 cents or so. If the kid liked us, we were "numba one" but if we attempted to cut into his profit margin, he may call us "numba ten." Most of the time, if the block of ice was large enough, you could lay the can on the block of ice and begin to spin it and it would melt into the block as you rotated the can. In no time, you had a nice cold soda. So here we were, a soldier of the U.S. and Vietnam consumer all rolled up into one.

Occasionally we crossed paths with some plantation workers who were gathering the sap from the trees. They used a knife designed to cut a groove into the bark of the smooth skinned tree. This groove circles the tree like the stripes on a barber's pole. At the bottom or end of this groove, a nail is hammered into the tree and a small bucket is hung to catch the sap as it slowly seeps down the groove of the tree. The workers move from one tree to another refreshing the cut on the tree and emptying the bucket of sap into a large vessel sitting in a cart powered by a water buffalo.

Every few hours we stopped to take a break from the weather and the tension that was in the air constantly. When you were tired, you normally just dropped down from where you were standing and hit the ground. As you took in the surroundings, your eyes automatically scanned the area for forms of life, any kind of life. At first the ground was void of any activity. Soon however, the dirt began to move with insects. How they knew we were there awaiting their interest is beyond me. Mostly what showed up were large black ants. They invaded your turf and were soon crawling all over your body. These types were harmless and were seeking food. They didn't bite; just investigated the large intruders occupying their world. I brushed them off and closed my eyes for a few minutes. Shortly, we got the word to move out and were on the prowl again as we continued the mission.

It was rare for our unit to be in the field with less than two platoons. This amounted to around 60 men if you counted the HHC. This is headquarters staff. HHC amounted to the C.O., his RTO, an F.O., forward observer for the artillery support, his RTO, and a medic. If we were to leave a basecamp via foot, then the size of our combat strength could be tailored to whatever was required to carry out the mission.

If we had a mission that required our unit to be placed into the middle of nowhere or where a surprise arrival was required, this meant that we would be transported via Hueys or slicks as we called them because they didn't have any external armament showing except for the M60's used by the door gunners. Flying via chopper became a logistics issue. At times we were limited by the number of air assets or serviceable choppers that were available. Most of the time we had around eight to ten and could put six men to a chopper. Doing the math, and we could put around 50 to 60 men on the ground on average with each airlift.

Did I mention that one of our weekly routines was to take quinine tablets to prevent us from getting malaria? It was the medic's responsibility to ensure everyone got the *white pill*. We also had some other tablets for water purification which we were supposed to drop into any water that we got other than out of our official water trailers. This was to keep us from contracting other bacterial diseases.

April 29th started at 0600 and time for everyone to get up, grab breakfast if available, or coffee and standby for orders. Our platoon leader, 1LT Chris Brown, had been summoned to a briefing at company HQ. Orders had come down and our platoon sergeant went over the operation for the day. We were to saddle up and join 1st and 2nd platoons to be airlifted to village AP 13 BIS, where we would set up a cordon and search the village and check it out for weapons caches and question any suspects we rounded up. They would know we were coming and what we were up to, so there won't be any surprised guests in the village unless we executed our mission with precision. Choppers would be arriving from the 187th AHC Aviation Battalion, *Crusaders,* at 0730 hrs from Tay Ninh. We were to be staged for pickup by 0715 hrs. The squad leaders returned and briefed the men. They picked out who will be going and who would be held back for the day. On this day, Sarge needed the full squad for fire support, so we quickly checked our gear. I looked over my magazines, grenades, both frag and smoke, and checked my canteens and filled them with water. I had cleaned my weapon last night, so once I got the order to *saddle up,* I was ready to go.

If we make contact, we may be out overnight. Although we had some breakfast, we might find ourselves enjoying a cold meal later in the day. Resupply would have to send out some food if it's called for; a pallet of C-rations. This would be done by a CH-47 if the size of the supply drop was large enough or it could be done by a Huey if we needed just a few cases of C-rations. Once the boxes were dropped off, the platoon sergeant would hand out the rations. To avoid everyone picking out the good meals and leaving the rest, the case of C-rations was flipped upside down before opening so the labels faced down. The boxes were then scrambled so you never knew which meal you would get. Selection of a box was given to the troops first, then the sergeants would pick last, always putting rank (private before sergeant) aside for morale reasons.

Although everyone called them C-rations, that term was incorrect. The Army started producing MCI's, meal, combat, individual rations starting in 1958 until 1980 when it was replaced by the MRE, meal, ready to eat. The MCI consisted of a rectangular cardboard box containing one small flat can, one large can and two small cans. It had an M unit (meat-based entrée), a B unit (bread item) composed of the Cracker and Candy Can, the flat Spread Can (cheese or peanut butter), and a D unit (dessert). It came with a brown foil accessory pack and a white plastic spoon in clear plastic wrap.

The M units had twelve basic varieties grouped in three menus and four different entrees:

M-1 entrees were: Beefsteak, Chicken or Turkey loaf, Chopped Ham and Eggs, Ham slices, Tuna fish.

M-2: Meat Chunks w/ Beans in Tomato Sauce, Ham & Lima (also known as Ham and Mother*uckers) Beans, Beef Slices w/ Potatoes & Gravy, Beans w/ Frankfurter Chunks. M-2A had Spaghetti w/ Meatballs in Tomato Sauce.

M-3: Beef in Spiced Sauce, Boned Chicken or Turkey, Chicken w/ Noodles in Broth, Pork Steak in Juices. M-3A Meat Loaf.

The B (Bread) units came in three varieties:

B-1: Crackers and two chocolate discs (solid, crème, or coconut) and Peanut Butter Spread.

B-2: Hardtack Biscuits with Cheese Spread (pimentos or caraway seeds) or Cheddar Cheese.

B-3: Cookies and a packet of Cocoa powder or Jam (Apple, mixed berry, seedless blackberry, mixed fruit or strawberry).

The D (Dessert) units came in three types:

D-1: Fruit—Halved Apricots, Sliced Peaches, Quartered Pears, Fruit Cocktail; D-1A Applesauce.

D-2 Cake: Pound Cake, Fruitcake, or Cinnamon Nut Roll. D-2A Date Pudding or Orange Nut Roll.

D-3 Bread: White Bread.

Within 10 minutes we got the word to head over to the air strip.

I grabbed my gear, tossed on two belts (hundred rounds each) of 7.62 for the MG, grabbed my M16 and joined my buddies. As we stood out in the field waiting for that distinctive sound of slapping rotor blades against the morning sky we nervously puffed on our cigarettes. Soon we heard someone pop smoke. That was the normal routine for routing air traffic. The ground forces toss out a smoke grenade and the lead pilot identifies what color it is. That way he doesn't get decoyed by the enemy into landing somewhere where he gets shot down.

Smoke was out and soon we saw our eagle flight coming in low on the horizon. Within minutes, dirt and dust was flying and we were trying to protect our eyes from the flying debris. As the choppers sat down onto the ground, we climbed aboard and settled in for our flight. The engine was whining, boosted by its turbine thrust. The rotors created lots of noise, too loud to hear anyone unless you were talking into his ear. One by one the choppers lifted off the ground, as the pilot tilted the rotors and we began our ascent. As we began to pick up speed, the pilot dropped the nose of the chopper which brings the tail up, like a scorpion. Within a few seconds, he hauled back on the stick and we shot vertically upwards. Everyone kept their eyes on the rotor of the chopper in front of us until we were safely airborne and had arrived at our cruising altitude. Sometimes the pilots had a tendency to crowd the airspace of the ship in front of him. No one wanted to see a rotor striking another chopper. No one talked, each man immersed in his own thoughts. Will the LZ be hot? Will we see any action today?

I looked over at my comrades and smiled. Today will be a good day. I have to put my trust in my faith and my beliefs. I told myself when I left the world that I could not change the outcome, and to trust the Lord. SGT Hugh Bishop made with some light jokes as he tugged on a cigarette. He was never far from a beer and always seemed to find some kid to bring him one when we were out on patrol. Our eyes were constantly scanning the ground for booby traps or trip wires. The VC liked to build pits and put sharpened bamboo stakes called punji sticks coated with poison or human excrement which were set in the bottom. They covered the top and hoped some poor fool stepped into the pit and impaled his foot. It happened far too often and our boots, in spite of a steel plate in the bottom of the sole, did not always protect the foot. The VC tried and set the stakes so the side of the foot could be stuck since our boots had canvas panels for air circulation.

The VC used any type of explosive device that could handle a trip wire as another of their favorite toys. We had as many casualties from booby traps as we had from enemy engagements. Grenades and unexploded artillery shells ranked high on his list of resources to use.

Something else we had to be on the lookout for was the vast tunnel complexes that dotted the countryside. The Cu Chi area had massive amounts of them. They were small and inconspicuous. It is amazing how small a typical South Vietnamese male is. They can fit into these tunnels that the average American can't get into. The tunnels consisted of nothing more than a hiding place, to having rooms which included supplies and even hospital quarters. We found first aid supplies in one bunker underground that was wrapped in a Chicago newspaper.

We had *tunnel rats*, volunteers who would go down into these tunnels with a flashlight and a .45 cal pistol to see what was in there. It was a dangerous job because you never know what you would end up facing, the enemy or some other crawling creature. A lot of the time, if we were suspicious, we would toss a grenade down a hole and the concussion would usually kill anyone close by.

We heard the crackle of the radio and Brasher, our RTO gave us the thumbs up. Our LZ was coming up and we were told to get ready. We had an escort today. We had several friendlies from Tay

Ninh; two gunships which had received small arms fire from the tree line. We were told the LZ is hot and the choppers won't be touching down. We had to jump and hit the ground running. The gun ships were circling the LZ and as we approached, the Huey gunners were beginning to pepper the trees and brush lining the LZ with their M60's. This was deafening to my ears because the door gunners had taken the flash suppressors off the barrels and it increased the crack of each bullet as it left the barrel. The eagle flight swept in quickly and our chopper nearly hit the rear rotor of the chopper to our front. I swore silently and tightened the grip on my M16. "Go, go, go," yelled the door gunner and out the door we went. We spread out and hit the ground as we waited for the choppers to clear the LZ. Soon, it was quiet except for the gunships which continued to circle our location. Charlie was silent now. Maybe he got his kicks in and now had decided he was outnumbered and it was time to get back to his village.

We waited for the signal all is clear, and slowly rose up and got into formation. 1st platoon had point this day and we were in the rear. We sent out our flankers and took up our azimuth reading. We got the word to move out. Over the next hour, we set up around the village, blocking all exits, and began a sweep of the village. That day, we didn't find anything. We carefully checked ID's and mingled with the people. The smells of the village covered a wide range of fragrances. We had mud huts with water buffalo living side by side. There were pigs in pens, and chickens running amok. Water did not look fit to drink as it sat in wells and clay containers. Used water was randomly tossed in any direction. Here and there, you got the smell of someone cooking on a WOK. Pork is the mainstay meat along with white rice. The odors swung from pungent and distasteful to intriguing as we passed through the curling smoke of open oven fires.

Everyone chews beetle nut, except for the very young. That stuff turns your teeth black or what looks to be black. Mama-san looks like she is 60, but is years younger. It's a tough life. The younger women, ages 14-20, I guess, were very beautiful and small in stature, but seemed to age very quickly in this county. Most of the women and old men don't seem to mind our presence, but it was hard to find young men beyond those in their teens. I think they were either in the ARVN or were missing due to the war, or just maybe, were members of the opposition forces.

After spending most of the morning in the village, we pulled out. The kids were hawking their sodas and ice. They were always cheerful and trying to get something off of us. They didn't want food and what we did have, they were not interested in eating. If we were away from the villages, they found us on their bicycles. We had to chase them away because you could never tell when we would be shot at. We had orders to make a sweep through the countryside and rally at checkpoint F. It was there we were to be picked up by chopper at 1400 hrs and returned to Dau Tieng. It turned out to be a quiet afternoon and all went well. We arrived at our checkout 15 minutes early and took a break, waiting for our ride. Soon, the command to pop smoke was given, as we heard the distant whop, whop, whop from the Hueys as they approached our location. The smoke was out and everyone began to feel some relief from the tension which will quickly leave as we climb aboard the Hueys and lift off.

Upon arriving back at basecamp, we stored our gear, and waited for mail call. Today, we not only got news from the world, we got a *goody pack* which arrived every week or so containing cigarettes, candy, and writing supplies among other articles for personal care. The goody pack, as it was called, was something we all looked forward to getting. It was a constant thing during everyone's tour in Vietnam. We didn't have access to a store, or PX. There was no way for us to get normal items that people at home take for granted. Hungry, well what's in the refrigerator or what can I get if I hop in the car and go downtown? No, here we depended on the goody pack to deliver those miscellaneous needs of daily life, regardless of where someone lives.

"Mail call," Spoores announced. "Gather around, or cover for your buddy." He read off the names as he went through the pile of mail he was holding in his hand. I received my first letter from Mom and Dad. I opened the letter and got my first bad piece of news. Uncle Richard, Dad's brother, passed away days before I left for Vietnam on March 9th. I am saddened to read this message. He was always one of my favorite uncles.

The goody pack was delivered to every company and the contents then split up in equal shares for each platoon. In turn, the platoon sergeant (PSG) would divide the contents up again by squad. The word would go out by the PSG to the squad leader (SL) to pick up his share. The SL then called his squad together and dibs to what was in the box was done by rank and seniority. Usually, the sergeant of the squad would let his men choose first, then he would rummage through the leftovers.

The item in demand first was the choice of cigarettes, followed by candy and other items. Here is a list of the most common items in a goody pack. For cigarettes, Marlboro, Winston, Camels, Lucky Strikes, Pall Malls, Chesterfields, Salems and Wolf cigars with a plug of chewing tobacco and Prince Albert pipe tobacco. Not a bad selection there. Candy was varied with Juzyfruits, Good N Plenty, Lifesavers, Chuckles, Hershey's and Spearmint and Blackjack gum. The everyday need items of toothbrushes and paste, plus shavers, blades, shaving brushes with cream, writing utensils, pens, pencils and writing tablets, and lastly envelopes were always in demand. Other items like soap, matches, sewing kits and pipe cleaners were also in the box. Not exactly the department store, but it made life in the field more bearable.

Since I had been in country, I had been smoking Salem cigarettes and I needed the menthol to cool my throat. They took getting used to but it's all I could take right then. Brasher turned on his radio and we listened to some music from AFRTS, Armed Forces Radio and Television Service. At times they would broadcast the Wolfman Jack show out of Los Angeles. It was time to wind down and clear our minds. Our nerves were wearing thin and I for one was developing stomach problems from the tension.

The Army had an ability to change ones outlook on life. Where race and color were issues in civilian life, for the most part, they didn't exist here. Everyone was called by their last name. There are no brothers of racial groups for the most part, just buddies. We quickly developed relationships, but I would stop short of saying how close we would get. The old saying of here today and gone tomorrow made you cautious of getting too friendly. It was hard enough to lose friends at home. Here, in war, you needed to keep your focus and when someone was killed, you just moved on, or at least I did. I really shouldn't speak for others and how they handled it.

Soon forgotten was the idea that it would be a year before you saw home. Eventually, you stopped thinking about what was going on at home and what your friends were doing. There was a disconnect from the people stateside and the job you were doing for Uncle Sam. It seemed unreal at night to be listening to the radio from California and hearing good ole rock and roll. Here you sat, drawing a puff from your cigarette, sharing jokes and conversation about home life at night. But, as the sun rose, life took on a different meaning, survival. At night you could escape your surroundings and dream of times past.

I had been in country since March 13th and, for the life of me, I cannot recall the names of anyone in my platoon outside of Price, Hill, Zimmerman and Spoores. The company got more replacements throughout the month. First platoon got a new platoon leader by the name of 1LT Louis Greika. Eddie Wales, who would become one of my best friends later, arrived in country in April, but he was assigned to D company at that time. Later around May, he would be transferred to Charlie Co. Ken Christianson and Felix Santisteven from Los Angles, California, Darrell Kuhanu, Minnesota, Lawrence McCloud from Greenville, Mississippi, Richard Polus, Indiana, Don Bonner, Hoxie, Arkansas, Richard Skogin from Alabama and others joined Charlie Company, a total of 18 in all. As these new faces arrived into each platoon, their personalities, behaviors, and the process of getting to know each individual could impact some of the chemistry that existed within each squad. Are they a head case or maybe a wise guy or are they just a normal kid who will fit into the role he is assigned without upsetting everyone around him? Time would tell.

8. VILLAGE LIFE

Out and away from any large cities, the villages of Tay Ninh and Dau Tieng were two of the four largest enclaves in our area of operation. The other two were Trang Bang and Cu Chi, being the largest of all. The pace of life was slow and easy for these farmers. Yes, there were other trades or roles, village chiefs, provincial heads, a few teachers here and there.

The villages varied in size and shapes, usually determined by high ground that rose above the monsoon rains that arrived beginning in May and increased in intensity until they begin to fade away by September. The indigenous people worked hard to supply themselves with food and a small amount of money they earned by selling livestock, mostly ducks, chickens, and pigs, or fresh vegetables that they took to market. The average farmer only made about $40 U.S. per year.

There were three styles of houses, brick, thatch and a combination of thatch and mud, in the rural areas where we conducted operations. Prior to the wars that had ravaged these lands, were red brick colored construction block built homes with window and door openings, but no windows or doors, ceramic tile floors, and tile roofs. The construction timeline when these homes were built is unknown, and most of them outside of large cities had been destroyed. They may have been built prior to WWII when the lands were peaceful and prosperity abounded. You found them around the villages of Trang Bang and Cu Chi. You could say, they were more like middle to upper income housing simply because of the materials they were built with. The mud hooch or grass hut was typical of most lower income homes and were built from materials of the land, nothing manufactured. They had straw thatched roofs made from rice plants left over from harvesting. The base of the home had mud and straw mixed together to form outer walls about three feet high and six inches thick. The rest of the wall was open construction. The floors were hard packed mud as were the sleeping quarters where a grass mat would be rolled out to lie on. The hooch varied in dimensions from 10 by 10 ft. and slightly larger. Each hut had a cooking area, and heat was provided by a wood fire. The poles holding up the roof were used as places to hang containers or cooking utensils. Many hooches had spider holes somewhere built into the floor for concealment protection. There was a lack of any furniture, maybe a few chairs, other than hammocks. Tradition for these people is to squat down, not sit. With a bit of practice we learned to mimic them and you could actually rest in these positions, but most of us did not make a habit of it. The trick was to really spread your knees and drop your hind end as low as possible so you were resting almost flat footed and your hands on the ground. The third style of house was constructed with tile or corrugated metal roofs, but could have a thatch roof. The walls were made from rows of stitched or tied together thatch material from the harvested rice which overlapped each other and did a decent job of keeping the rains out of the interior. The entire building was enclosed from the weather and some had doors and window coverings. Most of these homes were single story.

In other regions, you may find some two story dwellings, all using a variety of construction materials described previously. These houses usually have the livestock on the bottom floor. This is where the four or five year old would be tugging at the rope that is anchored to the nose ring of a water buffalo which weighs anywhere between 600-1200 lb as he leads these animals out to the rice paddies to plow the soil in preparation for planting season. The buffalo are kept in a sturdy cage of heavy bamboo poles and railings built into the bottom floor of the hooch. There may be other livestock there as well, chickens or swine. The aroma of this barn like environment mixes with the other odors of rotting vegetation, wet soil, and nearby latrines. It's a very unpleasant mixture to be forced into the lungs.

There were numerous trees in and around these villages to provide food and shelter from the sun. Banana trees grew everywhere here and are actually a plant, not a tree. The fruit is starchy and low in sugar content if not ripe but can be eaten raw. Their large leaves provide for a good sun blocker and also are used to wrap food in while cooking. There were Jackfruit, large light green fruit with a rough textured thick skin. We never figured out when to pick them when they were ripe. There were some pineapple plantations near us, but we were never there at harvest time. Most fruits found in the local markets seemed to be hiding in the jungles and areas we traversed and what we did find in the field in Vietnam were not to our liking. They lacked flavor, moisture, and appeal, or once again, maybe our ignorance of their harvest times kept us from sampling them.

Every village had a *dinh*, a communal house or temple with a pond in front and a shade tree in the back and often a Buddhist pagoda or shrine. The dinh, is where the village elders met and a guardian spirit, usually associated with a Vietnamese hero—resided. Every year festivals were held to celebrate this spirit. The dinh was a combination of the temple and the community center in many villages. It is within the dinh that the housewives offer prayers not said at home. At such times, the *thanh hoang* (spirit ghost of someone who died a violent death) was asked for protection against the various natural disasters and for his good will toward the individual worshiper or the worshiper's family.

The thanh hoang could also have died from an unnatural death such as murder, childbirth, or failed to be buried; or a supernatural or celestial spirit without human origin. Though the villager may claim his faith as Buddhism, Confucianism, or another of the ten or so faiths in Vietnam, the animistic belief of spirits who can affect and control destiny is very strong. Often the courtyard of the Dinh or adjoining temple had a lotus pond with the large round green leaves floating on the water's surface. The lovely flowers of the several varieties of lotus rising above the dirty water gave color to the area. They reminded the beholder that, as the beautiful flower grew in such a humble environment, so good could come from each regardless of surrounding conditions. (Source: The religions of South Vietnam in Faith and Fact).

Clothing is simple for these people. Most men and women wore a white silk shirt and black silk pajama bottoms. It was not uncommon to see both men and women drop their pants and do their business right where they worked in the fields. They did it so quickly that your eye could not catch or glimpse anything. Underwear was unheard of, except for the ladies who may use a type of wrap which acts as a bra. The younger women in the cities wore more elaborate tops, white pajama bottoms then had a colored top and shell of silk fabric that was split down one side or both and was worn over the bottom white pajamas. It was tied around the waist with a cloth belt. Everyone wore some type of basic sandal held on by straps or a flip flop shoe. Most everyone wore a conical hat made from bamboo to protect them from the sun.

Schooling was not available in the poorer areas that we operated in. I cannot recall even seeing a school, even near the larger villages that we occasionally walked through. I'm not sure how intentional this was or simply a fact that the country was so far behind other parts of the world when it came to educational practices.

Medical care was scarce for these families and when the time made itself available, the Division would make humanitarian efforts to win over the local population. Occasionally a MEDCAP would be scheduled, also referred to as a civil action program. The term stands for Medical Civic Assistance Program. We would go out to a village somewhere where we were not on good terms. The village would be secured by us, the infantry, by setting up a perimeter around the village. Inside the village, our medical staff, the Battalion Surgeon, his assistants and staff of other doctors if we had them, plus our medics from the battalion aid station and our own company medics would join forces.

The families would receive treatment by the medical staff for all types of ailments. There were infected wounds and sores, various diseases that were treatable, usually a large dusting for lice and other common bugs that took up residency in clothing and hair. The staff would issue vaccines and shots where needed. The kids loved this type of attention and the mama-sans appreciated getting their children doctored up. This process gave us a chance to note what the composition of the village was too. Most often there was a large part of the male population missing, that is, the younger men. They were either off in the ARVN army or serving the other side.

The MEDCAP would last until the supplies ran out or we ran out of patients. Everyone ended up happy and hopefully, there was a bit more trust generated between us and the villagers. For many, this was their only chance at receiving true medical care.

The infrastructure of the land was very crude. The exception was the major cities, like Saigon, Nha Trang, Da Nang, etc., that had paved streets, water, and electricity. All the rest of the country roads were dirt, red dirt, and were almost impossible to navigate in the winter. They turned into mud. Anyone living outside of these major cities relied on oil lamps for lighting, dug their own wells for water, and used some form of outhouse for sewage if at all. There was no form of communications such as telephone service or radio.

Transportation was by foot, bicycle, moped, motorcycle, and the Vespa van. These vans were heavily overloaded with passengers and market goods and wound their way from small hamlets to villages like Dau Tieng, Tay Ninh, Go Dau Ha, Trang Bang, and Cu Chi, and on toward the capital center of Saigon. The blue smoke from the overworked motors casts its mark over the roadways of the country. Cars were rare, but do exist, although not many of them were very modern and up to date.

Over time, the U.S. Army Engineers worked feverishly to update the main roads by reinforcing them with crushed rock and later asphalt to make them all weather roads and to minimize the enemy's ability to mine the roads. It was time consuming to spend each morning walking with the battalion engineers as they swept the roads for signs of mines, then were stuck with the task of clearing them without endangering themselves. On occasion they missed one, and, unfortunately over time, one of our rigs, be it a truck, Jeep, or tank, would find it and detonate it.

They had a mixture of religions they worshiped. First is the Vietnamese folk religion or basically the non-religious population which comprises 82% of the land. In Tay Ninh is the Cao Dai religion (Caodaism 1%) or triple religion and Temple. *Cao Dai* literally means the highest Lord or Highest Power. The symbol of the faith is the Left Eye of God. This religion practices the teachings of three faiths combined; Confucianism, Buddhism, and Taoism. Next, there is a large following of Buddhists (8%) and then the French introduced the lands to Christianity and the Catholic Church (7%) followed by Protestantism (7.5%) and Hoahaoism (1.5%). (Source: Wikipedia)

As the U.S. became more heavily involved with the war, the President of South Vietnam, Ngo Dinh Diem became a trouble point for the U.S. He was not helping the cause to unite the people against North Vietnam by inciting a personal vendetta against the Buddhist population. Diem was raised a Catholic and was a leader of the Catholic element in the country and opposed by the Buddhists. In protest against Diem's stance on religion, there were demonstrations by the Buddhists in public and many monks set themselves on fire in protest of their treatment. This was going on in the early 1960's. It was suggested by high government figures from South Vietnam, and with the blessing of the CIA, that the South stage a coup d'état. In November 1963, they assassinated Ngo Dinh Diem and his brother Ngo Dinh Dhu by Nguyen Van Nhung, the aide of the leader of the ARVN (Army of the Republic of Vietnam), General Duong Van Minh. Diem was replaced with Nguyen Van Thieu, who served as South Vietnam's second president until the country was defeated by the North in 1975.

He fled to Taiwan, then later to London, England. In the early 1990's he moved to Foxboro, Mass. when he lived in seclusion and died in 2001. He never wrote an autobiography or any papers about his past or the troubles of his home country. His aversion to public appearances was attributed to a fear of hostility from South Vietnamese people who believed that he failed them.

9. TRANG BANG AND FULLBACK 6

May signaled the start of the rainy season that would last through September. May 1st and 2nd were spent patrolling the rubber plantation to the east but there was no action. May 3rd we were south of Camp Rainier near the Saigon River and got the opportunity to take out a sampan. With so many waterways in this country, moving man and materials could be done by boat. I think it was around this time we went out on an ambush patrol. It was an uneventful mission. Early the next morning, in my haste to get my gear on and move out with the platoon, I left my wallet lying on the ground. I discovered my mistake when I got back to basecamp. I knew it was unlikely I would ever see it again. I quickly wrote home to tell Mom and Dad what took place. Shortly after that happened, I got a letter from the State of California, Department of Motor Vehicles. Prior to going into the service, I had accumulated some traffic tickets in my '63 Corvette hardtop convertible. The police just loved me to death and took every opportunity to spotlight me and give me a citation. Some I deserved to get, I can't deny that. So, back to my situation in Vietnam. I got this letter from the state asking me to forfeit and surrender my driver's license to them because of too many points against my record. I really had a good snicker over this one and wrote them back, saying that if they wanted my driver's license they were welcome to come over here and look for it because I had lost it and was too busy fighting the VC to go looking for it. Anyway, by the time I returned to the States, my license suspension was over.

The next day, May 4th, Charlie Co was put on a 30 minute alert status. We headed to the airfield and soon caught a ride via Chinook CH-47's down to Trang Bang to secure an artillery battery and provide security to a laager site. Charlie Company was OPCON'd to 1st Brigade while the rest of the battalion remained in Dau Tieng. We were placed under the operational control of the 2/22nd (M), call sign Fullback Six. They were operating to the northwest and south of Trang Bang. We were told there were approximately eight VC battalions operating in the area. That night as we settled into our new home, we experienced a mortar attack.

The evening of May 5th we were introduced again to the accuracy of the VC's skills. For the next few days we were holed up in a larger site on the western edge of Trang Bang, a small village at the crossroads of several major supply routes. During the night of May 7th, around 0300 hrs we were hit with 49 rounds of 82mm mortar and the 2/22nd (M) suffered 10 WIA's, while we had 4 WIA's. No one was seriously injured requiring a Dustoff. It was from here that we slowly moved eastward between the dates of May 6th to May 10th. Our time was spent sharing our location with 2/22nd HHC and running RIF's south of the MSR that cut south of the village of Trang Bang.

From Mike Smith, A Co 2/22nd: "Two 81mm gun tracks and our FDC track arrived at a culvert in the main road east of Trang Bang at dusk 5/5/68. There was a 'Dragon wagon' tank hauler (M88 Tank Retriever) parked north of the culvert off the road. I now assume it was the C 2/12th we hooked up with. We put an APC to the west, one to the east and FDC command in the center of the perimeter. It was raining when we arrived, and shortly after, the hasty bunker constructed by the 2/12th collapsed on their command group, RTO, CO, and two others. I was on the .50 cal to the east, so I was not able to be part of the quick excavation. Wet sandbags can kill. The dust-off came in at dark and lifted them off. Our 1LT Sorgee was in command of C team Bravo, as 6. Weird shit, the dust-off is gone then a car came up the road, stopped, was confronted; he needed gas. I told him all we have is diesel. Seems like a nice guy beating curfew and low on fuel. He had a candle on the dashboard next to a Virgin Mary. I am Methodist and think, good Catholic man, our side. Retrospect: The tank hauler trailer was there to bridge the culvert if they over ran us and blew it. The NVA sighted in or confirmed sighting (their mortar tubes) on that dashboard candle. When it started, they were around the entire perimeter. Their 75mm Recoilless rifle first hit the gas tank of the big truck and back lit the field. It was a mess. It was a coordinated attack across the A.O. We fired all our 81's in an arc to the west. In the process we set a rice hull stack on fire, it back lit them, but limited the vision of dust-off and resupply choppers. Choppers came in and resupplied all. We ain't talking shit, we fired over 100 rounds and were resupplied and fired 50 rounds of that. I grabbed bandoleers of M16 and went hole to hole to the legs. It did not stop. The sweep of the area in the morning found the hulls of at least 50 rounds of 75mm. We had to sweep the road to Trang Bang. Then the legs got on and we went to the village east. There was a split in the road. The ARVN fort was at the point. They had about 80 NVA KIA's in a pile out front of the wire, but they would not come out and sweep the town. I was the pivot man and the 2/12 spread out. We are one man about every 10 meters but this is a town. I step in the house closest to the fort. It was Vietnam NY. NY. All glazed tile floors. Problem they kept their prize water buffalo in the same family space. The people left, the NVA killed the animals, so no cattle a' mooing, and I was sliding through clotted blood on glassed tile. We got to the back of the town, mounted up and went to another...."

On the 11th, our company was split up into four units along with A and C companies of the 2/22nd (M) Infantry. Out of this we were made into three maneuvering teams comprised of M113 APC's and infantry. There were four platoons in each team, 16 for example represents 1st platoon; the 6 represents the platoon leader. His call sign is Charlie 16 and his RTO is Charlie 16 X-ray.

Team A (TM A): A Co 2/22 16, 26, 46; C Co 2/12 26;

Team B (TM B): A Co 2/22 26, C Co 2/22 36; C Co 2/12 16, 46;

Team C (TM C): C Co 2/22 16, 26, 46; C Co 2/12 36;

The team's objectives were to sweep the MSR daily, and alternate routes between Hoc Mon and Go Dau Ha daily. To sweep means provide security for the engineers who are using metal detectors to find mines or look for visible disturbances in the road surface and investigate them. Once this was done, the roads were opened to civilian traffic. After the road sweeps were completed, we moved off road to perform RIF's of the areas that parallel the MSR. Each team had their own section of road to sweep. Team A was located between Go Dau Ha near Gia Binh on Hwy 1. Team B was at the midpoint between Trang Bang and Cu Chi near road TL-7A while Team C was spending most of its time at night down near the junction of Hwy 1 and Hwy 237 just above Hoc Mon. From here, each morning we departed our night larger around 0600 hrs and left for our assigned checkpoints, escorting men from the 65th Engineers who were performing their mine sweeping duties, then proceed to conducting RIF's where we could maneuver or escorted resupply convoys down to Duc Hoa from Cu Chi.

There was a lot of enemy contact during this period. Most of the elements that made up the 1st Brigade were in contact daily. The organic units assigned at the time were the 2/22nd (M) *Triple Deuce*,

3/22nd *Regulars*, 4/23rd (M) *Tomahawks*, 2/14th *Golden Dragons*, 4/9th *Manchu's*, and 3/11th ACR (Armored Calvary Regiment) known as *Black Horse*. Joining us for a short period as being OP-CON'd to the 1st Brigade was the 1/5th (M) *Bobcats*. There were a number of local villages which seemed to be VC strongholds, X. Rung Cay (2), AP Bau Dieu and X. An Duc along Hwy 1. There were daily firefights engaging the enemy, but throughout this time our Teams were not impacted by the skirmishes. We remained in this configuration until May 26th when we were released back to the 3rd brigade.

The smell of diesel fumes was very strong. One of those manly wartime fragrances we were privileged to enjoy. The smell of diesel today takes me back to riding in the back of a deuce and a half or on the *tracks* of the mechanized units which was where I now found myself.

Some of the area we were operating in did not bode well for track units. They needed some room to operate and preferred some open space so they could maximize their maneuvering speed and leverage the firepower from the .50 cal machine guns mounted on the tracks. The first lesson learned about working with tracks was to never ride in them, but ride on top of them.

They were thin skinned with only 2" aluminum armor plating on the sides. An RPG will cut right through the sides of a track like a hot knife through butter and rattle around inside, thus killing or injuring everyone inside. So, the mantra was, ride on top and take your chances, then hopping off at the first sign of trouble. Another sidebar to track life, was avoiding red ant colonies. It was hilarious to watch, but painful to experience. The red ants were rather large in size, maybe a ½" in length, or slightly less. If you took a match stick from a paper match book and jabbed it at a red ant, they would grab it with their pinchers. You could get into a tug of war with them. It's impressive to feel their strength in this contest.

The ant had some incredible strength. They normally lived in nests which were like a hornet's nest in trees. When an APC broke through hedgerows, sometimes they knocked these nests down onto the track and within seconds, the vehicle was quickly abandoned, and troops were stripping their clothes off trying to eradicate themselves of the ants biting into their flesh. It was very painful, but once they were removed, your skin was ok. This was such a common occurrence that the 2/22nd made up an award called the Order of the Red Ant that was given out to guys who experienced this confrontation.

The area we were operating in with the *tracks* was outside of the rubber plantations that dominate the north. South of Go Dau Ha, Trang Bang and the area that is toward Cu Chi, on either side of the main supply routes is farming country. Tracked vehicles were only effective during the dry season. Trying to use them in the open rice fields was ineffective or impossible during the rainy season when the earth was too saturated with rain to traverse using heavy armored vehicles. The areas east and south of Trang Bang toward Hoc Mon were NVA and VC traveled routes. These were also staging areas and strongholds for troops and supplies. We foraged around and through these areas daily, hoping to catch the enemy off guard. I liked working with the mechanized units. There was a sense of strength and power in all that metallic hardware.

The terrain off the main supply routes was wide open rice fields. Some of these fields could be as small as one acre and others as large as 500. Mixed in with these open spaces were hedgerows of trees and brush that act as natural fence lines. Many of the small villages or hamlets were generally found in clusters of trees and bushes that were surrounded by open spaces and were on slightly higher ground, above the flood plain of the monsoon season. These were the areas in which we found concentrations of troops, supplies, and fortifications. It was never a good thing to go on a combat assault, and to exit the aircraft standing in wide open space with no protection and having the enemy firing at you and you had no cover. The same held true operating with the mech units.

We were just getting settled in with our new friends and were setting up our checkpoint south of Hoc Mon off to the side of one of the main supply routes. This would be our night laager site that we would use throughout the month. I did not see our resupply guys stop by our laager site to drop off the mail. It was around 1700 hrs and Richard Polus and William Potts in unison sounded "mail call" and I got several letters. They were from Susie, Patty, and Pam. I found a spot to sit, resting on the rear gate of an APC and proceeded to read my letters, and took a few minutes to fire back some responses, including a letter to Mom asking for some writing materials, and some banana nut bread.

Life with the 2/22nd was different. At night, we set up in a large circle with each track (APC) facing outward, equally spaced apart to provide a field of fire that covered the full 360 degrees. In the middle of the circle was the command and commo tracks, the HQ portion of our small task force. The commo track had a variety of whip antenna's on board which linked the unit to Battalion, artillery, and other units operating nearby. If we were alone, we would be digging foxholes every night for protection, but here we somehow felt a bit more secure. It did not alleviate the concern we each shared about where we would go if we got mortared. As long as the ground was hard as concrete, it made digging any fortifications to protect ourselves a monumental task. We made do the best we could.

The Mech boys had it made in some regards. They could get inside their tracks to get out of the weather and hot sun. It's also like dragging around a large storage bin. They had places to store their C-rations, extra ammo, chairs, radios, and other creature comforts. The down side is they were a big target for the enemy to go after.

One could characterize an APC as a large bee. If you started messing around with one, sooner or later you could get stung by the .50 cal machine gun they pack. The impact of these bullets was enough to saw down trees. Some tracks were also equipped with a flame thrower and were referred to as flame tracks. We didn't run across one of those too often.

We're all in this war together and aside from the typical jabbing that goes on, as who was better, mounted troops (mechanized) or grunts, the true foot soldier, we knew we needed each other to perform our jobs. We learned the ins and outs of working with the tracks and how we were supposed to fit in tactically with them. It's more than just catching a ride on top of one.

FIREFIGHT OF MAY 13TH AND 14TH, 1968

We had been with the 2/22nd just over a week. At first we were providing security for HHC 2/22nd. Soon however that changed and we were brought to the party. Glancing over the battalion daily journal logs, the 1st Brigade was making contact every day. The area we were operating in ran from northwest of Trang Bang to Go Dau Ha, south to the north fringe of Hoc Mon, south to Duc Hoa and again to the northern side of Cu Chi and had been a hotbed of activity. Every day, we were being sniped at, along with RPG fire. The NL's were being hammered with enemy 82mm fire. Granted, not everyone was experiencing that, but at least one unit in the operation every night did.

May 12th was an uneventful day with everyone completing their missions with little fanfare. However, that night, and into the morning of May 13th, much changed. Beginning with a mortar attack on Alpha 46, 2/22nd NL position followed by RPG and small arms fire that lasted for several hours from 0020 until 0230 hrs. The battalion journals were very unclear where this team was located on the May 12th journal. Earlier in the day, they were way down near Hoc Mon. However, the journal shows A46 2/22 reporting 3 KIA's and 4 WIA's, location CP8. At 0505 hrs the casualty report was updated to show 4 KIA's, 11 WIA's, one 2 1/2T truck destroyed, 2 APC's hit by RPG's, damage unknown. One low boy, a total loss, and one gunship hit. By 0235 hrs, the request for two Medevacs was complete. Silence fell upon the NL site.

The most prominent landmark in the entire region of III Corp was Nui Ba Den, or the Black Virgin Mountain. It rose about 3000 ft. above the flat terrain that surrounds it. It was a vital communications location for the 25th Infantry Division's radio traffic. There was a security force there supporting the signal corps that operated the equipment and maintained all the radio antennas. There was a number of small buildings, mess hall, communications bunkers, a helo pad, and a scattering of firing bunkers on the top of the mountain. The VC maintained control of the bottom of the mountain and it had been difficult to root them out. To go along with the rest of the attacks happening all over the region that early morning, May 13th, an attack on Nui Ba Den kicked off 'the Battle of Tay Ninh.' On top of Nui Ba Den, 140 men, which made up the security team and radio relay staff was overrun by a platoon sized VC force who nearly wiped them out before being beaten back. Spooky, the Gatling gun equipped C-47 gunship was called in to support, but the clouds covering the mountain top made that difficult to accomplish. Special Forces arrived to help the local command regain control of the situation. A day later, Alpha Co, 2/12th was rushed to the mountain top via Chinook helicopter to reinforce the security team.

While the 2/22nd was under attack, four ARVN compounds were also receiving incoming fire. They were located about three clicks SW of CP8. Further to the south, the 4/23rd was under mortar attack and to the East, the 2/14th was being tested too, all between the hours of midnight and 0300 hrs. Back to A 2/22nd's NL site, after receiving mortar fire, at daylight they proceeded on their road sweep and upon completing that mission, swept north to 500 meters above CP5 and turned around and moved south, arriving above CP8 around 0900 hrs. Upon the completion of their engineer and road sweep, they moved on to investigate the area of contact that they were in earlier that morning. As they approached the village, they began to receive AW and RPG's from the village of AP Bau Dieu, just off Hwy 1, halfway between Trang Bang and Cu Chi. They maneuvered to the west side of the village and returned fire.

The fight continued through the morning hours and the 2/22nd had 9 WIA's, some requiring a Dustoff. In the melee of the fight, SFC Charles H Sandberg, from Philadelphia, PA., 44th Infantry Platoon (scout dogs) was killed along with his dog, Buskshot (not a typo error). The mech unit tried to sweep the village, but the resistance was strong. 2/22nd radioed for a LFT and CS ship to assist. By 0930 hrs, a request for a Dustoff to remove six WIA's was issued. An APC was hit by an RPG. Clearance was granted for artillery fire which was soon pelting the village where the VC were taking shelter.

My unit, TM C was called in to assist in the fight, taking up a blocking position to the southeast of the village and began to stop and collect people vacating the area. A unit of 65 ARVN's arrived as well and started to interrogate the people being held. VC were observed leaving the village to the NE and NW. While this was going on, the 2/22nd sent in two platoons to sweep the village. At 1320 hrs another request for a Dustoff was made for three WIA's. It was 1600 hrs and TM B arrived to support TM A on their right flank on the SW side of the village. What was confusing, was the activities of TM B. They may have been assigned to stay in their larger site at CP31 on this day since there was no location information in the daily log until they responded to A 2/22nd contact in the afternoon. The location of CP31 could not be established from the daily journals.

The LFT from 25th Aviation, Diamondhead out of Cu Chi continued to circle overhead and targeted several hooches where the VC were firing AW's from. At 1845 hrs, the first of three airlifts arrived north of the village. C 2/14th arrived to secure the NL at CP8. Shortly following, B 2/14th combat assaulted the north side of the village, taking fire as they landed; three slicks were hit, one bad enough that it had to be airlifted to Cu Chi for repairs. B 2/14th was in a blocking position and now the village was completely surrounded.

It was 1655 hrs and the 2/22nd was receiving some 82mm fire. They reported their situation at 1745 as three KIA's and 26 WIA's U.S. and 16 VC BC (body count). There was still some sporadic fire coming from the village. Fullback 6 informed Viking 6 (2/14th call sign) we were up against at least two VC companies and possibly one battalion of VC. It was dark now and each unit sent out ambush patrols. All was quiet, for now. At 2222 hrs, Cu Chi and 2/22nd HQ location came under mortar attack. 2/22nd HQ reported and requested a Dustoff for one litter and six ambulatory wounded. At the same time the mortar attacks occurred, Nui Ba Den radio relay station reported that they were under heavy attack from an unknown sized force. They were receiving AW, SA's, mortars, and rockets. They were engaging with all organic weapons on station. A LFT and Spooky gunship were put on standby. Some bunkers had been destroyed. The enemy had infiltrated a few spots in the perimeter. S-2 (intelligence) had information that there was a 1,000 man force heading toward Duc Hoa to conduct an attack on that outpost. We were tripping over the enemy movement every time we broke out of our laager sites.

The sun was rising on May 14th and we had held our positions during the night. The time was around 0800 and two foot patrols, one each from A 2/22nd and C 2/22nd were ordered into the village to check out the situation. Within a few minutes after talking to the local villagers, word was radioed back to command that the VC had left the area. Fullback 6 told the whole operation to prepare for a full sweep of the area. A 2/22nd and C 2/22nd set up roadblocks on all roads leading out of the village to check all oxcarts and villagers for weapons and possible VC suspects. As the area was swept, the network crackled with word of two VC bodies found in one area and three in another, bringing the body count up to 23 VC killed. At 1130 hrs, another two bodies were found and the sweep continued. As the teams moved through the area, one APC from C 2/22nd hit a contact mine, leaving a 3'x5' crater. There were four WIA's and the APC was a combat loss.

Mid-morning, back at Dau Tieng, two platoons from A Co were air lifted by CH-47's to the top of Nui Ba Den to reinforce the radio relay station after yesterday's assault by the VC which killed 24 soldiers, mostly from the 125th Signal Bn and from the 3/22nd and 4/9th which was acting as a security force.

By 1500, it was all over and the operation was abandoned. The sweep had confirmed 34 VC KIA's. All three teams headed to their NL. Orders were given for each team to send out AP's for the night. A 2/22nd 36 was assigned to security for a 3/22nd convoy headed from Trang Bang to Cu Chi. The day was finally over.

Elsewhere in the area on May 13th, elements from the 4/23rd (M) and D 2/14th came under heavy attack or made contact with large enemy forces. S2 (intelligence) reported several large forces were seen moving south and southeast of Hwy1. One group of 1000 was preparing to attack Duc Hoa. D 2/14th engaged in a company sized force near XT751166. At the same time, the 4/23rd was having problems of their own. They suffered four KIA's and 20 WIA's. They found a few bodies with NVA uniforms on. It seemed that Charlie had been seen wearing green uniforms, black pajamas, a sprinkling of NVA garb, and a few were killed wearing ARVN outfits. The 1/5th Bobcats were OPCON'd to TF Daems (named for LTC Leonard Daems, 3rd Bde Commander, Jan to June '68), at 1200 hrs but returned to 1st Brigade command around 1855 hrs. Their casualty report stated one KIA and 14 WIA's, but their action location must have been under someone else's log report. They set up a NL in support of the 2/22nd at 1900 hrs.

Those two villages on Hwy 1 between Cu Chi and Trang Bang, X. Rung Cay (2) and Ap Bau Dieu, remained a festering sore as long as the *Warriors* operated in the area. May 17th, the VC mounted another attack on units operating in the area of these same villages. The VC opened up with AW, .51 cal, and RPG on TM B. Guys from C 2/12th 1st Plt, which made up part of TM B, SGT Michael Granum and his squad, including Elmer Lightner and other members of the platoon, Jessie Anderson, Leon Barnett, Freddie Chism, Steve Divan, Felix Santisteven, Richard Scogin, James Simone, and Craig Schoonderwoerd in 1st Plt who were with the 2/22nd team, were seeing a lot of combat and found themselves in another village fight. RTO Billy Zimmerman was busy on the radio. After attempting to slug it out, the team withdrew to safety and called in artillery strikes. The team attempted to reenter the village and once again were receiving fire. Again, artillery and air strikes pounded the village. As the team withdrew a second time, PFC Gerald Le Blanc, from Louisiana, who had been with the company only a short time, was killed by sniper fire. There were three WIA's reported. By 1700 hrs, the VC broke off contact.

Elmer Lightner later said, "I can't shed a lot of light on Le Blanc. If I remember right, he was only there a couple weeks or so and I really didn't get to know him. I believe he was hit by small arms fire as we were trying to set up a defensive position after breaking contact with Charlie. It was raining and we were working with some APC's and one round hit the perimeter. We were not far from a small village and a cemetery but I don't know which one it was."

The 2/14th, 3/17th (M) and 4/23rd (M) found themselves going head to head with large VC forces east of Hoc Mon and put a hurt on the enemy, killing 17 VC.

May 23rd arrived and like all the previous days, the story remained the same. It was during these types of missions that it was easy to drop your guard and become careless. The NCO's reminded us the enemy can strike at anytime and anywhere.

That morning, TM A continued to operate split apart with the Mech heading north from their NL to clear the MSR and the Infantry (2/12th) sweeping south on a RIF from their NL at FSB Stuart, located on the southeast edge of town on HWY 1. Early in the morning at 0900, an ARVN compound reported receiving mortar fire and SA fire from an estimated force of 50 VC. 4/23rd was dispatched to sweep the area around 0940. At 1400 hrs, a Little Bear (25th Aviation BN) chopper reported receiving fire in the same area. After sweeping the MSR road and outposts, TM A (TM A—MECH) was ordered to return to the same area and make a sweep. At 1440 they reported from their location they were receiving RPG and small arms fire. They had one WIA and one APC hit by RPG and was non-operational. The team returned fire with organic weapons and requested LTF and FAC with air support. At 1445 TM A MECH reported no longer receiving fire. TM A had pulled back to the MSR and had requested a CS ship. They were told to outpost MSR and to send a second convoy thru vicinity of contact. LFT was on station and TM A was no longer receiving fire. No casualties were reported and no body count found.

In the south, TM B—INF (C 2/12th 1st / 4th Plts) completed their RIF. They ended up at their NL of TM B around 1200. At

1215, TM B was assigned a bushmaster (ambush) at a trail crossing in a hedgerow area. At 1225, six VC walked into the area and were engaged with Claymore mines, the VC dropped to the ground. Negative fire was returned. After waiting, at 1300 TM B swept the area and found several blood trails leading to a bunker then departed to TM B NL site. TM A at 1420 ran into VC who were located in the hedgerow laced with spider holes and they opened up on TM A—INF. The team pulled back while receiving RPG, small arms, and AW fire. They returned fire with organic weapons then withdrew, making a request for a LFT by 1LT Ford. The earlier request for a dog team arrived by Huey at TM A's location. The LFT, 25th Avn Bde, Diamondhead, was on station. Smoke was popped to ID their location.

The gunship was given identifying marks where to fire at the enemy positions. We withdrew to safety while the dog teams sought shelter in a hedgerow. The gunship made a pass in error at the hedgerow where the 2nd Plt and one dog team was located and fired 6-8 rockets, hitting the G.I.'s. The result was 1 KIA (SSG Billy Parrish, Visual Tracker) who took a direct hit. Dog handler Rodney Marrufo ran to assist Parrish and was hit by another rocket, killing his dog and severely wounded himself, along with four WIA's. Among them were Abney, Woodson, and medic Andy Wahrenbrock who was sent home with a broken ball socket joint from the impact. At 1625 TM B—INF left NL in route to area of contact and arrived to support TM A.

38th IPSD (Infantry Platoon Scout Dog) dog handler Jerry Suitor (attached to HHC 1/27th) helped load some of the WIA's onto the gunship, including Doc Wahrenbrock after this event. SSG Billy Parrish was dead, missing limbs. The platoon withdrew and the KIA (Parrish) had been left in the area of contact. At 1835, TM A reported they were firing into VC positions and had sent a probe into area to retrieve the KIA. Upon retrieving the body, they started receiving mortar and RPG fire. They pulled back again and were counter firing with mortar and artillery fire. They had requested a Dustoff to pick up the body. At 1840, TM A reported they were returning to NL location. Total report was one KIA, five WIA's and one VC KBA (killed by artillery fire). SP4 Rodney Marrufo died at 12th Evac Hospital. Later that night at 2030 hrs, TM A reported that one of their five WIA's had D.O.W, now two KIA's and four WIA's.

Dewitt Roberts, who was with the 38th IPSD, Scout Dog team said, "It was a dark day for members of the entire 66th IPCT (Infantry Platoon Combat Trackers) the evening SSG Billy Joe Parrish, and SP4 Rodney Marrufo were KIA. Jerry Suitor (scout dog tracker) and another unknown 25th Div. Grunt loaded SSG Parrish on the Gunship that killed them. Parrish was blown in half at the waist by a chopper rocket with the exception of a bit of skin. They had him folded in half in one of their shirts used as a stretcher. Suitor was in shock then and his circuit breaker appeared to have spared him some of the details to this date."

Andy Wahrenbrock recounted his time at 12th Evac after the action, "There I was, on a gurney heading for skin graph surgery, location the 12th Med Evac unit in Cu Chi, South Vietnam, in 1968. I vividly remember that morning being wheeled to pre-op down a hallway while IV bottles on the shelves rattled from the concussion of B-52 five hundred- and thousand- pounders dropped along the wire of the 25th Infantry's sprawling basecamp."

***

Other units working away from the MSR's were finding the VC daily. During the month of May 1968, the enemy attempted to move large units of troops south out of their refuges near the Cambodian border. They infiltrated southeast of Dau Tieng and down through the Iron Triangle, crossing the Saigon River, then holding up in the Hobo Woods or Filhol rubber plantation before crossing Hwy 1 on their way to their objectives of Duc Hoa, Hoc Mon, and on to Saigon where they hoped to disrupt the civilian government and local way of life.

Our mission, aside from keeping the MSR's clear and operational, was to support the other units operating in the 1st Brigade and their mission of intercepting these troop movements. May proved to be very successful for the 25th ID and near the end of the month, during the last week; most VC and NVA units began to withdraw back to the north. They were under severe pressure and suffered some fairly high casualties attempting to overrun the ARVN outpost or whenever they attempted to engage our ground forces directly in daylight.

Arriving back at our usual NL, gave everyone time to wind down from the tension of the day. Price and Hill came looking for me to chat a bit about what went down during the day. We had a few light hearted laughs. Price talked about his home in Kentucky, a small farm in the rolling hills called Loretto and located in the middle of the state about 30 miles from Elizabethtown. He said he was *home born* and this was his first experience being away from home, courtesy of the U.S. Army. He's a handsome kid with striking features, and jet back hair. He has that natural southern slow drawl as he talks. He grew up in the woods and is a natural shooter.

William Hill, who went by the nickname of Fox, and I didn't know the origin of that, was from the Bronx in New York. All he had known was the big city life, but he was easy going and has a great smile. We have grown close over the past several months because we are all in the same squad. I tell them I need to get to my mail and want to send a few letters home. Mail call was announced and I got a letter from Mom and Dad, telling me that Grandma had passed away. She was

102. For the past ten years or so she suffered from dementia and at times did not know who Ernest, her son (my father) was. I think that was upsetting to my Dad. She was put in a nursing home for the last few years of her life. Uncle Richard had been taking care of her, but he just couldn't do it anymore. Richard was a bachelor all his life.

Charlie Company wrapped up its time with the 2/22nd with little fanfare. Some of the platoons got into a little action, but for us in third platoon, it was rather uneventful finding ourselves more in a supporting role than leading the attack. Maybe it was because we operated close to the division basecamp, or maybe we were just lucky. On May 26th, we were trucked back to Dau Tieng and our basecamp and returned to the control of 3rd brigade.

Upon arriving back in Dau Tieng, almost immediately, we were OPCON'd out to Fullback 6. We were given operational orders that had us running night operations and setting up ambush patrols around the area to our south and west, out in the Ben Cui Rubber Plantation along the MSR from Tay Ninh to Dau Tieng.

INTELLIGENCE SUMMARY OF APRIL AND MAY'S BRIGADE OPERATIONS

Intelligence concerning the Michelin, Trapezoid, Mushroom, Boi Loi, Hobo, and Hoc Mon area was obtained from many venues. These areas were being used by VC/NVA units to train, resupply, and recuperate from TET attacks. The protection of these areas was vital to the mission of the local force and main force units occupying them. The enemy engaged U.S. units in major attacks in night laager positions to prevent 3rd and 1st Brigade from moving into secure areas. The units in contact were elements of both the 271st, 272nd and 273rd VC Regiments.

Throughout the operation the enemy continued to attack night laager positions with ground and fire attacks. They continually engaged mechanized units with sniper and RPG fire.

The enemy fought delaying actions against U.S. units entering the major basecamps. These tactics allowed a number of the enemy to escape contact, though large amounts of supplies were found. Elements of the brigade moved to Hoc Mon area where VC/NVA units were attacking. Contact was then initiated by U.S. units in sweeps of areas. The enemy was trying to avoid involvement until reaching their objectives. In the area of Hoc Mon, the 1st Brigade was using four task forces to maintain control of the MSRs while initiating sweeps of the area.

Somewhere around May 28th, I got word that 1LT Brown wanted to see me. I located him at HQ and reported in. "Krause, how would you like to become my RTO?" he asked.

There was no hesitation from me. I replied, "When do I start, Sir?"

Brown responded, "My RTO is leaving. His DEROS date is here and I need someone now. You can start tomorrow. Head over and find Thompson. He'll give you a rundown on what to do. I'll let Spoores know what's happening. You can check in with me in the morning."

With that being said, I told LT Brown thanks and headed out to locate Thompson. I was checked out on operating the Prick 25, the nickname for the AN/PRC-25 radio. The radio was attached to a backpack frame and weighs in just fewer than 20 lbs. It boasted a recommended range of five clicks with the short tape antenna, up to eight clicks with the seven section long-range (fish-pole) antenna and was powered by a dry cell battery that had an average life of 20 hours. One of the tricks we learned about later was using the discharged batteries to operate our personal radios back at basecamp. The whole time that I packed the radio, I used the shorter tape antenna, material similar to a measuring tape. Most of the HHC, such as the company commander and the forward observers for tactical air support or artillery support, used the whip antennas because they needed the greater range to talk.

May 30th, all Charlie Co platoons were ordered to assemble in formation for a change of command and awards ceremony. The men had their fatigue shirts dressed with their pistol belts and everyone had their pants *bloused*, and steel pots on, standing at parade rest. Our BN CMDR, LTC Don Green presented 1LT Jay Hickey with a Bronze Star then had Hickey hand over the colors to his replacement. Hickey was being reassigned to HHC, a normal step for someone who was getting "short" in country while 1LT Ford leaves 2nd Plt to become company X.O. 1LT Ronald Hendricks formerly from Delta Co where he was a PL, takes over the company. Hiram Marziano, Hickey's RTO was asked by him if he wanted to work in company HQ, which he said "yes" to, and eventually became the company's NCOIC, handling all the formal paperwork, reports, recommendations for promotions and awards etc. It was a good safe job to have. Years later when Hiram and I talked about Hickey, as we were trying to locate where he's living now. He said, "I just loved Hickey. Jay loved to call in air strikes and bomb the hell out of everything or blow it up with artillery fire. That was great."

Several days before this, the company added two officers, 1LT's Richard Wiggins and R.W. "Bud" McDaniel. Wiggins took over second platoon from Ford while Bud McDaniel was temporarily assigned to HQ. 1LT R.W. McDaniel was an OCS graduate from North Carolina. Wiggins and McDaniel were good friends, having gone through OCS and jungle training in Panama together. Who would have thought after all of that, they would end up in the same company in Vietnam? Later that day, there was a memorial service led by the officers including LTC Green, to remember those men that were lost in the TET offensive and in the April firefight. It was a somber 30 minutes for everyone gathered and reminded us of the fragile situation we were in.

Platoon sergeants lasted far longer than platoon leaders. In my tour of duty, I had six platoon leaders but only three platoon sergeants. If they were RA's then they were usually E-7's. Why? Who knows? So here I am, no longer fresh and wet behind the ears, now following 1LT Brown like a small dog. But that was my job. I was always to be within arm's length of him in case he was contacted via the PRC-25. There was a constant chatter of dialog going on between the C.O. and the platoons. I didn't have a channel up to eaves drop with battalion. That was not my role. RTO's were looked up to, almost like another NCO. They needed to assess and transmit information as they saw it, many times without prompting from their officer, like giving a visual commentary of events.

May 31st, an excerpt from a letter I sent home:

> *To fill you in on something new, I have a different job now. Would you believe a radio telephone operator? Yes, a good old RTO. At least it's one more step to another promotion, but that will probably be another month or two yet. Your care package arrived safely and everyone enjoyed your cookies. I'd say they lasted about 15 min. But I got my fill so I thought I'd bring a little joy to the rest of the fellows.*
>
> *Yesterday I attended communion. It was a nice service and also was a memorial service for some of the boys. Seems like every month we have one. Death is so near sometimes. Day before yesterday a boy was killed by a mine and we had to go get the body. It almost made me sick. But you can't afford to let it get you down. All you have is to keep going on faith and a sharp memory of home and your family.*

Oh, I just remembered there was such a wild confusion getting back here to Dau Tieng. I lost those pictures and some letters. I was keeping them dry in an ammo can and the guy I gave it to, to hang onto it for me left it there in Trang Bang. Seems like I cannot hang onto anything lately.

The rains picked up and we were moving deeper into the monsoon season. The storms built up during the daytime and generally by midafternoon we got a downpour that lasted about an hour. Night time operations became more hazardous and more nerve racking. The flashes of lightning, along with a solid downpour, masked the vision and sounds, but it worked both ways. Trying to stay comfortable out on ambush was not easy, but part of the job. The light danced off some of those who wore ponchos to shield themselves from the rains. That made me nervous, and them a target. I preferred to tough it out and keep to my fatigues, as wet as I may get.

During May, our replacements were few and far between. Ronald Stepsie from Boyertown, Pennsylvania, Richard Conlin, a shake and bake E5, also from Pennsylvania, PSG John Partee, from North Carolina arrived. In the coming months I would get to know all these guys really well. More faces were likely to change soon. We had squad leaders that were getting near to the end of their tours.

10. TACTICS LEARNED AND USED

While we were on ambush assignment, other companies in the battalion were using some different tricks. Some other tactics that were being applied daily involved our approach to searching areas and villages. First, was the *area cordon and search*. This method was used when a large area or several villages were to be searched. The blocking ambush forces were positioned at one end and/or the flanks of the target area. The sweeping forces then moved through and searched the villages in order to push the enemy into the blocking ambush positions which had been established between the enemy and his sanctuary. At the same time the sweeping forces conducted a thorough search of the area. In order for this operation to be successful, the blocking ambush positions had to be placed between the suspected enemy and his sanctuary or escape route; and they must move into position without being detected. Two or more rifle companies were normally employed in this operation.

The second tactic was a normal village sweep. This was used when units moved through a village conducting a search while not in conjunction with blocking forces. This was the least productive and least desirable method of searching a village. This method was used both as a show of force and as an intelligence gathering media. The sweeping forces looked for large quantities of rice, bunkers, and positions, or indications of recent enemy use. We had been forced to use this method many times, primarily because of the non-availability of troops. If a unit conducting this type of operation could obtain the use of aerial reconnaissance aircraft to work in conjunction with their sweep, the chance of success was greatly increased. Village search operations should be conducted in cooperation with the local Vietnamese authorities, usually popular forces, also called militia or at times, the National Police if we were closer to metropolitan areas. Once the people were rounded up, they could be questioned by the Vietnamese personnel. We found that the National Police were ideally suited and trained for these population control activities when available. They were easier and less hostile with the locals.

As I learned, there were multiple types of terrain in which we fought. There were two wars in Vietnam; one was being fought in the coastal lowlands and valley plains, and the other one was being fought in the mountainous jungles. In some areas, units would be operating in the first environment one day and in the second the following day. The normal methods of operation differed sharply in each environment. Jungle fighting was not new to U.S. soldiers nor did the enemy have a monopoly on jungle know-how. U.S. units adapted well to jungle fighting.

When we were operating against the NVA along the Cambodian border, we found that they had as much difficulty operating in the area as we did. The prisoners we captured were, as a rule, undernourished, emaciated, and sick with malaria. They stated that almost everyone in their units had malaria and many had died from it.

In the jungle, landing zones were few and far between, trails were few and narrow. Navigation was difficult, units in many cases were limited to jungle trails and flank security was difficult to attain. Using trails was hazardous because we could be ambushed easily and tried to avoid them as best we could. Visibility was usually between 20 to 30 meters and forward movement was generally limited to 300 to 500 meters per hour. The most difficult problem in fighting NVA in this type of terrain was finding him. This was where he built his fortified basecamps and where he located his bunkers on ridges and in the heads of draws in hopes that a platoon or company would blunder into the area. The NVA habitually emplaced their fighting positions to fire down the valley or ridge; as in the fortified village, the enemy realized that our tactical advantage was in our artillery and air support. So, again he liked to use *hugging* tactics. Therefore, the problem was finding and fixing the enemy without having our units engaged and shot up at close range.

In the jungle where LZ's were limited, reaction time was reduced to the cross-country proximity of units on the ground. Commanders must continually consider the possible requirement to rapidly reinforce small units which gained heavy contact with the enemy. When operating in this type of terrain where contact with large NVA units was possible, rifle companies should be within zero to three hours of each other. Coming to the aid of a fellow unit under fire, at times, meant a cross country forced march as transportation by air was not always available.

The rifle company should operate as a unit with platoons within 15 to 30 minutes of each other. The company should have security elements covering the main body, front, flanks, and rear. Too often company commanders overlooked flank and rear security because of the difficult terrain. There were times when it is impossible to have flank security because of the heavy jungle vegetation. In this case, the unit moved in a single file. The point element would precede the main body by about 100 meters. A rifle company would stop every so often and send out patrols in all directions. Not only was this a good security measure, but it was also a good method of search in the jungle.

Special emphasis had to be paid to the rear. On the border, the NVA had developed a habit of once they located a U.S. unit, they would have a small recon party follow to keep tabs. There were a couple of ways to combat this technique; one was by dropping off a small ambush patrol. This procedure had paid dividends for us on several occasions. The other method was by having a patrol *button hook*, move off the trail and double back at some distance and move back along the trail. When operating in mountainous terrain, if at all possible a company commander should keep one or two platoons on the high ground, so that if need be they could maneuver down upon the enemy. That procedure paid off several times for our battalion.

Another part of Operation Toan Thang II, which started in June, 1968, involved disruption via ambush. In a linear ambush, security should be emplaced at least 50 meters up and down the trail from the kill zone and rear security beyond hand grenade range. Many platoon leaders were reluctant to put this security out because they felt they were endangering their men. The fact was, if he failed to put it out, he was endangering his entire patrol. The security was to provide early warning, and one procedure we had used was commo wire strung from the security to the main body; upon approach of the enemy, the security simply tugged the wire. The security would allow the enemy to pass his position and get into the kill zone. He would cut down any enemy that attempted to flee the ambush.

Using a Claymore mine with the security for the purpose of hitting the enemy as they fled worked well. Another problem was people falling asleep. No matter how much rest the men had received, some of them were going to doze. Commo wire strung along the position and wrapped around each man's wrist was one way to alert them of the enemy's approach. Here again, leadership was the answer. The patrol members would stay awake if the patrol leader made them stay awake.

Many times ambushes were properly planned, well laid, and correctly positioned, only to fail because of some small failure on the part of the unit commander; such as:

Springing the ambush early before the enemy got into the "killing zone."

Poor noise discipline, talking, shifting positions, slamming of weapon bolts.

Lack of sufficient firepower in the initial "springing of the ambush."

Failure to have escape routes covered by Claymores and or by artillery fires.

Failure to provide for illumination in conjunction with springing the ambush and with a sweep of the area.

The division considered the ambush, particularly at night, one of our primary weapons against the enemy. There were several additional types of ambushes that we deployed.

1. Stay Behind Ambushes

In some areas the VC habitually followed friendly units keeping tabs on them and feeling that the safest area around was where friendly troops had just vacated. It was a good practice to occasionally drop-off a squad to set up a stay-behind ambush. The ambush element must have communications with the main body and the main body must not get so far ahead as to be unable to return and assist the ambush element if this became necessary.

2. Claymore ambush

One of the most effective weapons to employ in an ambush is Claymores. The ambush site must be selected to take maximum advantage of the Claymore's capabilities. For maximum effectiveness, the Claymores should be located approximately 5-10 meters from the trail. Each Claymore should be sighted in, to ensure a thorough coverage of the killing zone, and their fire fans should overlap. To ensure a simultaneous detonation, they can be rigged in a daisy chain using detonation cord. Also Claymores may be used covering escape routes out of the ambush area and to help provide flank and rear security to the ambush unit.

3. Use of Anti-Intrusion Devices

The anti-intrusion device with the trip wire area covered by Claymores can be an extremely effective small ambush. The trip wire should be spider-webbed across the trail, high enough so that small animals will not set it off. The intrusion device itself was with the triggerman a safe distance from the killing zone. Care should be taken in emplacing the device since the wire is fragile and breaks easily. The Claymores should be carefully sighted in and camouflaged. A Company rigs up this device for use by the rear security. In two nights, four NVA were killed by this method as they approached the ambush from the rear.

4. Trip Flare Ambush

One type of ambush that had some success in Vietnam was the trip-flare ambush. The site selected must be observed and within friendly artillery or mortar fire. Several days prior to preparing the site, artillery or mortar fires were adjusted into the target area. Then trip-flares were clandestinely set throughout the target area, preferably along trails habitually used by the VC. Then it was a matter of keeping the area observed. If and when a flare was tripped, artillery or mortars were immediately called in. The area selected can be within friendly small arms fire range and these weapons may also be used. The two types of ambushes used most successfully were the Linear Ambush and the Inverted L Ambush.

11. MONSOON IN THE RUBBER PLANTATIONS

June 1st marked the start of Operation Toan Thang II. Charlie Co was still under operational control with Fullback 6 of the 2/22nd. Our task had been night operations for the last week of May and we continued with this effort through June 10th, pulling AP's into the Ben Cui Rubber west of Dau Tieng. We would either walk out of basecamp or at times we would be convoyed to an area, wait for darkness, then move into our ambush sites. In the morning we would be trucked back to base.

During the first week of June, we were given orders that took the platoon or several squads out into the Ben Cui rubber plantation nightly where we set up along a road, ox trail, or path and waited for some unsuspecting VC to wander by. The second night we triggered two Claymore mines and killed several VC on bicycles. Were they truly VC or just some poor farmer violating curfew? We will never know. The bodies and possessions were searched for documents or other intelligence gathering information. We left the bodies where they lay, sprawled out on the road. Anything we found was turned over to the S-2 and our Kit Carson scout, usually a local ARVN soldier assigned to our unit. We also used Chieu Hoi's, defectors encouraged to leave the Viet Cong under the Open Arms program and help the South Vietnamese army. We dropped leaflets by air which are free passes to cross over to the ARVN Army. He could be a former NVA or VC soldier. Each company usually had one defector embedded into the unit like the Kit Carson. Their language skills plus knowledge of the local area was helpful, but some of us were reluctant to give total trust to these men. They had to prove where their allegiance truly lies and we could only answer that question by working with them and evaluating what they did and told us.

June 8th, it was starting to rain daily. Many of our ambushes at night were done in pouring down rains mixed with lightning storms. It was hard to try and sleep lying on the ground in water. I pulled my towel over me, covering my face from the rain and lay my head on my folded arm and tried to get some sleep. We took turns on watch, but nothing developed. As daylight approached, we were ordered to breakdown our ambush and return to basecamp. This routine repeated itself. Most of us felt sleep deprived. It was all part of the process. We got used to it and learned to sleep when we could. Some days when we were some distance from basecamp, the battalion would send out trucks to pick us up and ferry us back. This day, we had no luck and spent the early morning hours working our way back through the plantation, across the bridge spanning the Saigon River and into the outskirts of Dau Tieng village and finally through the north gate of Camp Rainier and back to our company area.

June 9th and our company had ambush patrol once again, but this time it would be different. Tonight, the company left the basecamp and split out into three different platoon sized ambush sites along a major supply route, Hwy 239, west of Dau Tieng village in the Ben Cui Rubber. Hwy 239 connected Tay Ninh to our west and the basecamp for the 1st Brigade, and Dau Tieng. We gave our equipment the once over and I did a brief radio check. I checked the PRC-25 to make sure I had the correct *push* or radio frequency for the day. I keyed the talk switch and start talking.

"6 X-ray this is 36 X-ray, do you copy?" releasing the transmit key.

"Roger 36 X-ray, this is 6 X-ray, I copy, over," the RTO for the C.O. barked back.

"Lima Charlie Hotel Mike, over," I called out. That means I hear you *loud* and *clear*, *how* about *me*?

"Roger, I hear you 6 by 6, out," the RTO responded.

"This is 36 X-ray, out." I set the *horn* next to the radio.

We were good to go. The company had orders to leave the "wire" (basecamp perimeter) at 1600 hrs. The azimuth would take us out into the rubber plantation. It would be dark soon and there was no moonlight.

We left the basecamp with 1st Plt taking the point followed by 2nd and 3rd Plts. I located 2LT Brown and fell in behind him in formation. The company moved to a checkpoint where we waited for darkness. Around 1700 hrs, just before we arrived at our holding point, we relieved elements from Romeo 6, Recon Plt, at checkpoint Golf. They returned to Dau Tieng basecamp. We arrived at our assembly area around 1830 hrs. The word was passed down from 1LT Ron Hendricks, our C.O. that it was time to head out to our AP's. It began to rain. I checked my watch and it was 1917 hrs. It had gotten so dark that everyone in the platoon had a hand tucked into the ammo belt of the guy in front of him. Without doing this, I couldn't see the guy to my front. My eyes would not adjust to the darkness. We were in the trees and there was no starlight. It continued to pour and all one could hear was the water splashing off the leaves of the trees and striking the ground. There was a roar to the downpour of rain. The water was collecting on the plantation floor. We got an occasional lighting flash that lit up the trees.

The night's operation had us setting up three separate ambush sites. 1st Plt dropped out of formation and moved off to the north to set up along Hwy 239 in the Ben Cui Rubber. Another 400 meters along our azimuth and it was 2nd Plt's turn to separate from the group.

We moved forward in the night, just a platoon. I was about 8-10 guys back of the guy on point and following 2LT Brown. I don't recall any of the names of the soldiers I was with. We moved in silence towards our AP. The platoon edged along in the rain, trying to navigate to where we were to set up the ambush. The wind was blowing along with a constant downpour as I wiped the water from my eyes and face. The company left basecamp close to an hour and a half ago.

There were occasional flashes of lightning again. Everyone scanned left and right keeping an eye out for anything moving. Another lighting flash cut across the sky and I think I see a group of people, five or six or more, dressed in dark clothing standing on a road off to our right front. The light was playing off their rain ponchos. Was my mind playing tricks on me? Did anyone else see what I did? I was unsure and we continued moving forward toward those figures. I was guessing they were about 100 feet away. I didn't say anything, unsure of what to do. The sky lit up again and this time there was no mistake.

Shots rang out and the platoon dived for cover. We were as surprised as Charlie was as we came face to face, less than 50 feet apart. Because of how we were moving, not everyone has a clear field of fire. I had a number of troops between me and the enemy. There was random and sporadic fire directed at the VC. They were returning fire as well. I could hear the crack of some AK-47's and then *Whoosh*. I knew an RPG was coming and there was a loud explosion to my right, then another one. This one sprayed me with shrapnel, nothing serious, just small fragments in the hands. Juan Antu and others were hit in the exchange. Finally, we were returning a base of fire at any muzzle flashes we could see and Charlie disappeared into the darkness and the firing stopped. The rain which was pouring down eased up but the wind continued to howl and blow hard. All this chaos ended and only lasted about a minute or two.

It was around 2100 hrs. We did a quick check to see what our casualties were. Juan Antu from Yvalde, Texas, who was in front of me and ahead of the Lieutenant, had taken a bullet. The round hit his helmet, spun around in the helmet liner and entered his skull in the back. I tried talking to him.

"Juan, can you hear me? Where are you hit?" I was getting no response.

The medic was busy with other wounded so I tried to see what I could do. I was feeling around on his chest and arms, trying to feel for blood. Even if I found some, how would I know with all the rain and our fatigues dripping wet? But, this did not stop my search for wounds. His helmet was off and he was lying in the wet mud on his back. I gently lifted him up to check his back side. He was limp and lifeless. I felt around on his head and it felt like a cracked egg. I tried to locate a pulse or a breath. Something, anything that would tell me he was still hanging in there, but no, I found nothing. He was gone. I radioed for an immediate Dustoff. The weather was bad and the winds and rain picked back up and were heavy and intense. We were told the Dustoff would have to wait until the weather lifts. Dau Tieng Dustoff says they can't go up in this wind. An hour later, Dustoff 77 attempted to brave the weather, then radios to us that they were aborting the mission and returned to basecamp. We waited and we couldn't get a Medevac for several hours from anywhere.

Finally, a crew from Cu Chi volunteered to come to get us, but the team from Dau Tieng made another attempt and braved the weather and arrived at 2240 hrs. I radioed to the chopper and tried to talk the chopper down to our location. We had a flash light we were using to get his attention. The light was placed into a helmet so it could only be seen from above. There was a clearing that we had moved to alongside the main supply road and a quick security perimeter was put in place. The chopper sent down in the small area we had flagged for his landing. We had more wounded than the chopper could take, somewhere between 10-13. It couldn't get off the ground. Someone needed to remain behind until a second Medevac arrived. We lightened the load. There were nine wounded on board heading back to Dau Tieng.

Our KIA was left behind and we had instructions to bring him back to camp in a vehicle in the morning. It had been a rough night. What was left of the ambush patrol waited near a main road for daylight and we swept the area once more. We found four VC ponchos, one with a lot of bullet holes, but no blood trails and no bodies. Not surprising in the heavy rain we had during the night. We reported a possible five VC body count anyway. As dawn arrived, a second chopper was ordered to our location, and then once again the flight was cancelled by battalion. Orders were finally given to come back to basecamp by convoy. We finally got Antu loaded on a deuce and a half and the entire company was trucked back to basecamp. We gave a briefing and afterwards, I along with three others, headed down to the aid station to get our wounds attended to. I was written up for my second Purple Heart. After leaving the aid station, I headed back to our hooch to catch some sleep. The final tally was 13 WIA's, four were hospitalized, and one KIA. One of the wounded was our commanding officer Ron Hendricks who remained with 3rd Plt as we set out on our ambush. He was hospitalized for three weeks and we got a temp C.O., 1LT Jimmy Ford, our X.O. to fill in.

I was getting the hang of being an RTO. It was a good job which I enjoyed and took great pride in. All communications were important and so was being accurate in detail, without being too wordy. June 11th while patrolling to the east we run into a squad of VC. LT Ford, who like Hickey, enjoys employing firepower, calls for an airstrike but gets little in return.

The next day, June 12th, one of Mom's care packages arrived from home. She was really good at sending items to me that I would like to have. For example she liked to send cookies, apples, dried fruit, and tea bags.

What was in my package this time? Yep, strawberry shortcake and it couldn't have arrived any later to be good. Along with the dessert, she sent me some salami which I quickly turned into some sandwiches, as least doing the best I could with what I had.

On the 13th and 14th, again we made contact and once more, Ford called for more air strikes. We were out in the rubber east of the basecamp. No one was hurt and we had negative results. Our platoon leader Chris Brown receives a reprimand and is sent packing to 3/22nd. In his place, 1LT R.W."Bud" McDaniel assumes the leadership of 3rd platoon.

The next day Charlie Co did a combat assault and was OPCON'd to the 1st Brigade, 101st Airborne where we laagered at FSB Allen, which was to the southeast of the base and was at the southern edge of the rubber plantation and at the top of the Iron Triangle. We stayed at this FSB until June 20th, basically providing security for the battery of 105mm guns and sweeping the area immediately around the FSB.

Alpha Co was released from the 11th Cav and OPCON'd to 3/4 Cav but remained in place. Brave Co was pulling AP's around Dau Tieng minus B26 (second platoon) which was up on top of Nui Ba Den acting as an additional security force. Delta company was conducting RIF's in the Michelin Rubber Plantation to the East, making contact with a small VC force, resulting in two WIA's and called for a LFT for support.

The same day that Charlie Co was airlifted back to DT on the

20th, Alpha Co was released from the control of the 3/4 Cav and also was flown back to DT. As Alpha Co got off the CH47's, Delta Co climbed on and took Alpha Co's place being OPCON'd to the 3/4 Cav in Trang Bang. Later that afternoon at FSB Stuart, around 4PM, PFC Marvin McCain Jr. from Delta Co's mortar platoon disappeared, and as far as anyone knows, that was the last time he was seen. A missing person report was filed and McCain who was characterized as being a loud mouth who spoke with a southern accent. He was 5'10", 160 lb, stocky build, brown eyes, black hair, from Alabama.

The missing person's report also said, "Last seen by members of his unit at 4:00PM near unit's perimeter. A short time later, muster was conducted for guard duty and evening men's and member's absence was noted. A ground search and aerial reconnaissance of area was negative." Delta company spends the next day, June 21st searching the area with aerial assistance and help from 3/4 Cav. He was not found.

Marvin remained missing until August 17, 1973. His remains were located and later, on November 9, 1973 identified as his. It is unknown where or how his body was discovered. At this point in the war, all U.S. ground troops had returned to the U.S. and only a handful of advisors remained in country. This story is just another oddity of the war.

Many years later, Mike Hauser had this to say about Marvin, "He was in my squad, 4th. The day he disappeared, he came to Spillner looking for some pot to smoke. He was dressed in South Vietnamese tiger stripe fatigues and was acting weird. One time in the jungle he kept getting lost so I tied commo wire to his belt and mine. He was a strange guy. Glad they finally found him."

On the 21st Charlie Co sent two platoons, 1st and 2nd to Nui Ba Den to reinforce the mountain top and replace Bravo Co. Craig Schoonderwoerd, Billy Zimmerman, and Elmer Lightner, part of 1st

Plt, reported that being on the mountain had a great view of the country, but it was freezing at night. All the Claymore mines had their wires cut by the VC attack and they got busy reestablishing the defensive positions around the perimeter of the camp.

"It was a mess when we got there. The wires to all the Claymore's were cut, so we had to fix those. The field of fire was poor because of the brush and rock formations. I just got a bunch of grenades and at night, would chuck a few down the hill just to keep the VC honest about trying to breach the bunker line again. We were up there for a awhile before we returned to Dau Tieng," Craig Schoonderwoerd recalls. They said they can't wait to get down off that mountain. While they were gone, we ran AP's until June 27th. The rest of the month was routine. We walked and sweated, searching for the enemy or attempting to catch him on the move using RIF's and then conducting nighttime ambushes.

The ambushes at night were scary, especially if it was raining. When it rains, it comes down in buckets and then you have to deal with the lightning and thunder. I was always thinking we were sitting ducks out there and could be seen for miles. As we moved to our ambush sites, everyone was tense until we stopped and got set up. The guys up front were scanning for any movement or dark shapes or silhouettes resembling a human shape. Everyone tried to move as quietly as possible avoiding stepping on any object that might snap or draw attention our way. Heads were turning left and right and everyone was straining through the darkness to see what might lie ahead.

As we arrived at our AP sight, we quickly set up the Claymore mines and checked the location of our weapons and located an escape route. We did our best to settle in and wait for the enemy to wander by. I took my turn on watch like everyone else, trying to maintain my focus through the sheets of rain and flashes of lightning. After two hours it was my turn to try and get some sleep. I lay down on the ground, knowing I would be lying in water and would remain wet all night long. I pulled a towel over my head to keep the rain from hitting my face and tucked my hands between my knees to keep them warm and closed my eyes. I used my helmet as a pillow.

Morning could not arrive soon enough. When it did come, and if we had not sprung our AP, we gathered up our mines and trip flares, made a quick equipment check and headed back to camp. Along the way, we dried out our clothing from the heat off our bodies and as the morning light got stronger, everyone was beginning to feel better about the day.

Weeks had gone by and the thoughts of when will I ever get back home had faded with time. No longer did I wake up in the morning with this thought on my mind. My mood and energy turned to sharpening my skills at survival. Like a football game when you first start playing, the action was too much to consume for the mind. But, as I gained experience and confidence, I saw events more slowly unfolding before me.

The rains continued to bathe our tired bodies day and night. We continued to go out at night on ambush patrols for the remainder of June, but found the effort proving to be uneventful.

I'd been an RTO for a month now. It was an enjoyable job because I could hear the order's filtering down from command and get a grasp of what the LT was trying to do as he evaluated situations and barked out orders to be relayed to the platoon. I had some sense of purpose and could see what we were doing tactically. In a way, I could anticipate what some of the orders would be and already be relaying them down the rank and file before they arrived, but not too quickly. Taking care of the radio did not take any time. My job was to make sure it functioned and that I had the operating pushes that we would be using with other units and our chain of command, and keeping the battery fresh. We kept bouncing around on the various bands so that we would not be intercepted by the enemy in case they were listening in on the net. The last few days of the month we continued to tramp around in the rubber plantations surrounding camp. The past few weeks contact had been sporadic. I think much of it had to do with the weather and the rain. The going was not easy when foot traffic had to navigate through muddy fields and standing water.

The month of June, 1968, was just plain crazy for the Battalion. During the second week alone, there were two reported incidents of accidental shootings by individuals at night along the berm line at Camp Rainier. Were they accidents or were these attempts by individuals to get off the line?

As time progressed, many of the items issued to me when I first arrived with the company were slowly disappearing. The first item to go was my gas mask, then later it was my poncho liner, a light cloth material in a camouflage pattern which was intended to be tied to a poncho to help keep you warm. The poncho was a rubber coated "rain coat" which had a hood in the center of what looked like a square piece of fabric. You tossed this thing over your head where the hood was, and the rest just draped over your body. The down side of wearing a poncho, was that it reflected light at night in the rain when there were lightning strikes and I felt like I was wearing a beacon for the enemy to see. I stopped wearing one because of this just prior to the June 9th ambush.

Craig Coleman from Fairfield, California, Lynwood Keatts from Gretna, Virginia, Carlos Serrano, Steve Ward, from Los Angeles, California, Percy Miller, from St Petersburg, Florida, Dale Freidig, from Washington arrived in the platoon. Jay Hickey, our former C.O., was also rotating home.

12. JULY IN HOC MON

On July 1st, I was promoted to Specialist 4th Class also referred to as Spec4 (SP4) for short. I didn't see the official orders until August. A SP4 is a grade above PFC, but has no authority to command troops. It put a few more dollars in my pocket. I had been sending money home to my Dad so he could make car payments for me.

I was not feeling well and came in out of the field ahead of the company on July 2nd and reported to the aid station. They sent me to 25th Medical. I had cramps and diarrhea. That was nothing new since I had been feeling that way since I got there. I was diagnosed with hookworm and they gave me *Kaopectate* and *Piperazine* for treatment, along with *Tetralac*. I went back to the aid station the next day and I was returned to duty. The company returned to basecamp on July 3rd.

From 1LT Richard Wiggins, age 27, 2nd platoon, "On the 4th of July, we had a unit party at our C Co CP area, and I remember the cooks along with the 1SG cooking steaks and we had a good bit of beer that day. Late that night, the VC attacked the airbase and we were the ready reaction unit and moved to the airfield to reinforce the bunker line there. We had a lot of incoming fire and when we did return to our GP mediums where we kept our belongings and slept when there, I remember thinking how lucky we were as there were many, many holes in the roof and sides of the tent from the mortars and /or rockets that hit close by. It wasn't long after that we left Dau Tieng and I went to take command of Alpha Co."

The VC swarmed the basecamp on July 4th at night after hitting us with hundreds of mortars. Rounds were exploding everywhere and the shrapnel was punching holes in the tree tops and landing in our company area, shredding our canvas tents. We were all out in the bunker line waiting the barrage out. Over on the other side of the camp, the VC attacked and we could hear the M60's and 16's lighting it up. The VC were hitting the wire with RPG's and getting in through the perimeter on the 3/22nd side of the base which was opposite of our encampment area. The VC were trying to hit the airfield and HQ area of the camp. Our company got a radio call and the LT got word down the line that we were to lighten up on the bunker coverage on our side of the perimeter.

We were pulled off the bunker line and moved into position near the airfield as a reactionary force in case the VC broke through the 3/22nd. But the *Regular's* held their own. The VC were repelled and thrown back. Some damage was done, but the VC paid the price for their bravery. There were bodies lying about in all shapes where they fell. Many still had satchel charges tied around their midsections. The next morning the area was swept and the bodies were collected and buried.

DEFENSIVE OF SAIGON — DEPLOYMENT OF 25TH DIV. WEST OF SAIGON, JULY 1968 — GIA DINH PROVINCE

July 5th and the word came down that the 1/27th, 4/9th and 1/5th would be OPCON'd to 3rd Brigade and dispatched down to Hoc Mon. From there they would come under the command of CMAC,

Capital Military Assistance Command, a component of MACV, Military Assistance Command, Vietnam, which conducted oversight for all military strategies and operations within the country. CMAC was responsible for operating the affairs of the military within the capital, including the coordination of its defense. Around July 8th, scuttlebutt had it that we were headed south too, to help rebuild a village destroyed by another operation. We heard that we would be there from one to five months. But, rumors are just that, rumors. But this time, they turned out to be accurate, at least the part about heading south to Hoc Mon area. SGT Dale Freidig gets the word from LT McDaniel that he's taking over 3rd squad.

My wrist watch was starting to act up. It took some of the shrapnel from that RPG hit back on June 9th and had scratches and indentation marks on the face. Watches were easy to come by. Like the soda kids, there were people out hawking watches and other items all the time when we are out patrolling. I needed something that would keep good time. The climate was hard on any watch with all the humidity and rain. Hopefully, on our next patrol I would find one.

On July 9th, beginning at daylight, Bravo Company was on the tarmac awaiting two Chinook CH-47 helicopters to take them down below Hoc Mon. This was the same area that the battalion was in during February and March right after the Tet offensive. It took two lifts per company. Our turn arrived at 0840 hrs and we were dropped at our LZ, XS758997. Joining us along a line of defense was the 4/9th, 2/22nd, 1/27th and the 1/5th. We were strung out over roughly ten miles or about 16 clicks along the "0" latitude map line between XS and XT. Basically, we set up a picket line, like a defensive fence tied at both ends with geography that was not navigable by large hostile forces. The way into the capital would have to be through us. We were assigned to the control of CMAC, in charge of the defense of Saigon. However, 3rd Brigade maintained operational field control.

Intelligence believed that the NVA were planning a mini TET and wanted to hit Saigon. Alpha and Bravo Companies were laager'd to our west at map grid XS740986, about three clicks away. Charlie Company joined Delta Company to set up a second laager site. We were in the province of Gia Dinh which runs from the coastline west and northwest, taking in Saigon and extending to an area near Hoc Mon. From our new laager site we could see the skies of the city lit up every night with aerial flares. The city was nervous after suffering from the surprise attack in February with the Tet Offensive. The bad thing about this situation, it impacted our night patrols. You could see people walking from great distances against the glowing horizon of Saigon. The landscape in this area was very flat and had immense fields of rice. We were in full swing with the wet monsoon season and this area gets around 8-12 inches of rain during the month.

The very next day, July 10th, we headed out to a village not far from our basecamp to perform an MEDCAP, Medical Civil Action Program. This was a humanitarian visit where we tended to the villager's health needs and treated everyone we could during the course of the day. It was an enjoyable time knowing we could do something for them that was beneficial. The children were always entertaining and would gather around us with the curiosity that only a child could exhibit. They wanted to wear our helmets and try and bum cigarettes from us or anything we had to eat other than ham and lima beans if we were carrying C-rations. This area is only five miles from the city limits of Saigon, so urban sprawl is prevalent and population density is heavy. Being here was a public relations effort to show we were there as allies, not enemies.

The VC were surprisingly quiet and were very crafty in picking what fights he wished to engage in. Being so wet, it was also difficult moving around and most activity for civilians was regulated to hard packed dirt roads and ox trails. We remained vigilant and kept off those travel routes for fear of stepping on booby traps or trip mines. Walking all day through fields of rice with your legs up to calf level, and sometimes deeper, in water and mud is exhausting. Another thing we had learned to live with were the water wells, which were everywhere. The local peasants dug large pits or holes in the ground which trap the rain water. That is fine as long as the area is not flooded with water. Many of these wells were also located in the middle of small berms which were built from dirt and were used to separate rice fields or paddies and acted as dykes. The wells were used to irrigate the land during the dry seasons, but in winter when it rained heavily, they got lost. You were walking along and all of a sudden you disappeared. If the well was not too deep you were ok, but these things could be 10-12 feet down or more. Loaded down with all the equipment we had on, it was easy to drown if your buddies couldn't locate you and pull you out quickly. There were a few guys in the platoon which just had a knack for falling into these wells, both wet and dry.

One of the enemy's favorite battlegrounds was the fortified village. They consisted of several hamlets which had been prepared with extensive fighting positions, trench works, connecting tunnels, and spider holes. The fighting bunkers often had five to seven feet of overhead cover and could take the near miss of a 155mm round. The bunkers were placed to cover avenues of approach into the village and were interspersed throughout the village to cover trails and approaches.

Many of the huts would have a fighting bunker in one corner. Tunnels connected the bunkers and trenches, allowing the enemy to disappear and appear firing from another location. Trees, shrubs and even the earth itself were reshaped to conceal those positions. At first glance, there seemed to be no logic or method to the defensive works. However, upon closer investigation, one found an intricate, well-planned defensive position that took advantage of the existing cover and concealment, natural barriers, and avenues of approach into and within the village.

The enemy elected to use a hamlet or a village as a battleground for several reasons:

1. He expected to inflict enough casualties on U.S. troops during the attack to justify his making a stand.
2. The U.S. soldier has a natural aversion to fire upon villages and populated areas.
3. The village offered the VC/NVA a labor source to prepare the fortifications.
4. In the open valleys and coastal lowlands the villages contain a great deal of natural cover and concealment.
5. The hamlets in a village were usually spread out and their arrangement offered many avenues of escape.

The enemy's normal plan of battle in a fortified village was as follows:

1. He would allow the U.S. troops to get as close as possible before opening fire, usually 15 to 25 meters. The purpose of these hugging tactics was to get the U.S. soldiers so closely engaged that they could not effectively use artillery and TAC Air.
2. The enemy felt that if he inflicted several casualties in his initial burst that our soldiers would become involved in trying to get the wounded back to the rear for evacuation. He believed that when the U.S. troops started worrying more about getting their wounded buddies to safety than about the battle, they were easy targets, and in that respect he was correct.

3. Another facet of his battle-plan was to fight viciously until dark, then, using the cover of darkness, escape using one of the many pre-planned escape routes, carrying off his dead and wounded, their weapons, and even empty cartridges. We had captured numerous enemy documents either condemning or commending certain units for their police of the battlefield.

On one occasion after an 18-hour battle, there was one particular bunker from which a LMG, light machine gun, was firing; after the fire fight and upon checking the position, not one empty cartridge case was found. After several battles, enemy dead were found; and lying by their side was a large tin can filled with empty cartridge casings. They were ready to move out when the signal was given!

The enemy knew that we placed great emphasis on body count and weapons. Our men, being typically American, expected to see at least 10 enemy bodies for every one of their buddies killed. The enemy knew that he had won a psychological victory if he could remove his casualties, leaving a sterile battlefield for our men to find, especially if he had inflicted some casualties on us. The enemy liked to initiate these actions in the late afternoon. That gave him several hours to inflict as many casualties as he could, then escape after dark. He did not have enough ammunition to conduct a sustained defense, nor could he be resupplied as our men could.

Therefore, if he began his battles two or three hours prior to darkness and held out until dark, he had an excellent chance to escape. In order to preclude the enemy from getting away, all escape routes must be sealed off. And this was indeed a difficult task; in fact it was usually beyond the capability of one rifle company. The impulsive company commander that attempted to use his platoons to maneuver and flank a fortified village soon found himself in deep trouble and the same was true if he tried a frontal assault. He may succeed in taking the position, but his losses would not be worth the attempt. His best course of action was to immediately call in blocking artillery fire to the rear of the position and utilize his unit to fix the enemy and give his commander an appreciation of the situation.

PHOTOGRAPHS

SGT Arnold Krause, on patrol, east of FSB Pershing, Nov '68

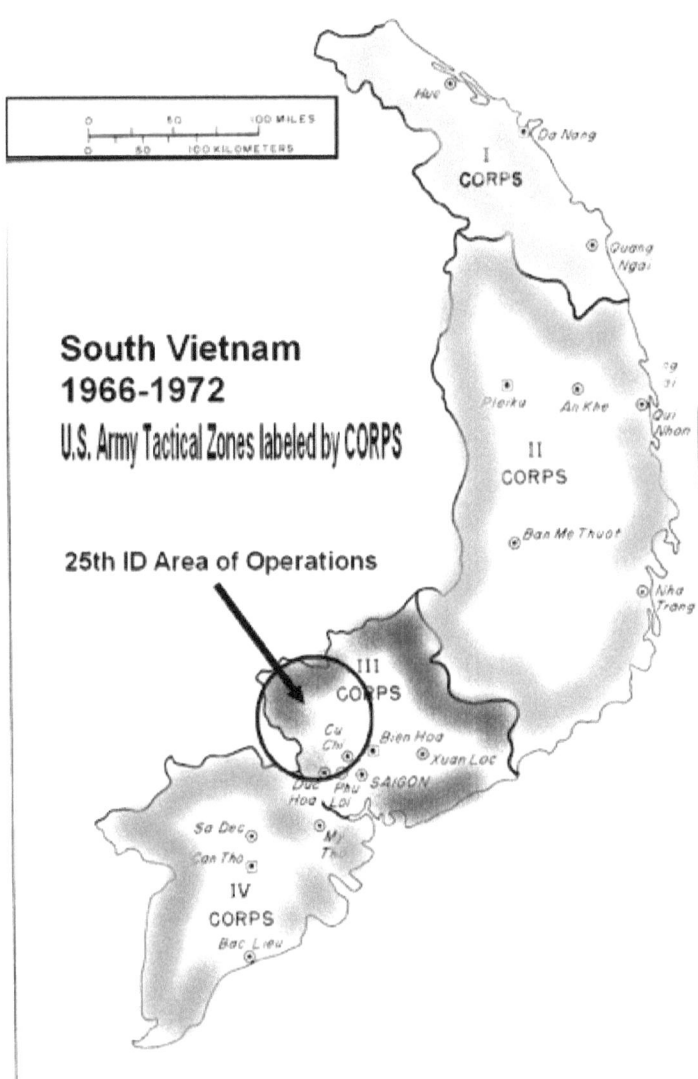

South Vietnam—Map of the "Four" Corp(s) Military zones

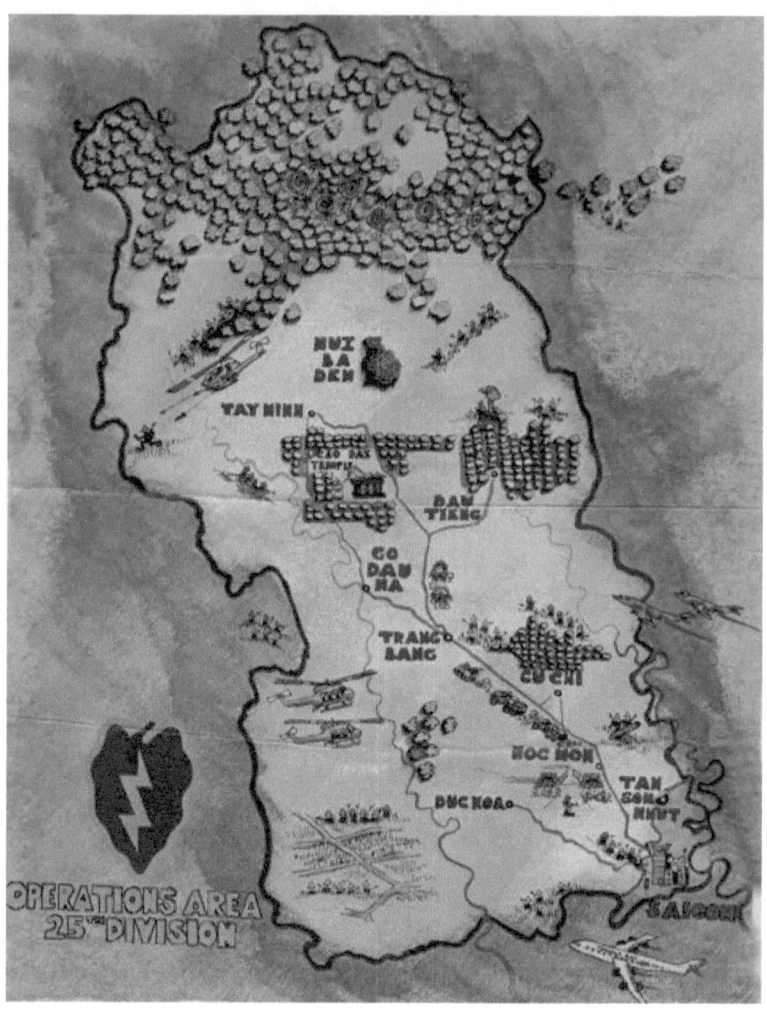

25th Infantry Division, Area of Operations in III Corp

Camp Rainier, Dau Tieng, basecamp of 3rd Bde, 25th ID

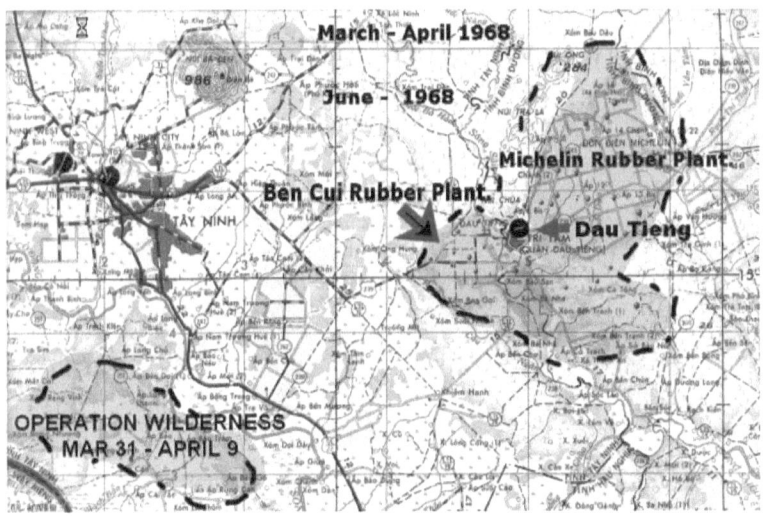

2/12th Map of AO—March to July '68

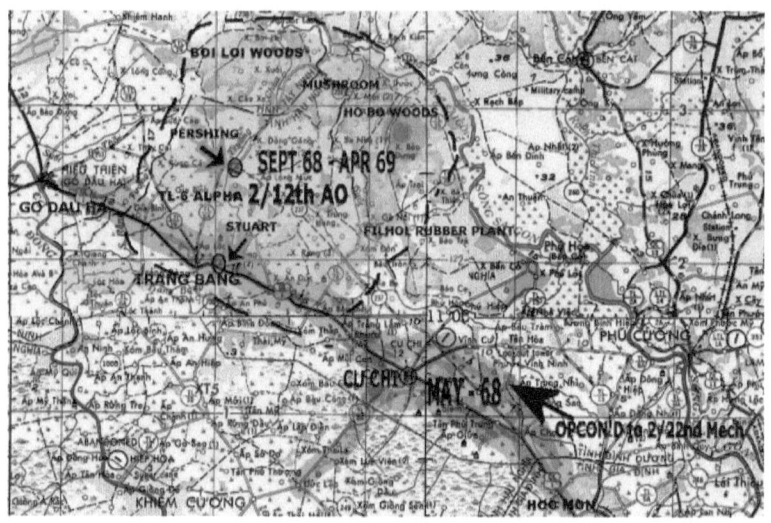

2/12th Map of AO — May '68 plus Sep '68 to Feb '70

1LT James Hawthorne's 1st plt, Dec '68 era, 2nd from bottom row, LT Hawthorne

1LT Jimmy Ford's 2nd plt, March '68, 2nd row fourth from left, LT Ford

1LT Mike Sheehan's 3rd plt, Jan '69 — Bottom Row — Nevers (resupply), Wales, 1LT Sheehan, Buckley, Keatts, Top Row — Satterswaithe, Hill, Price, Coleman, Christianson, Tostado

3rd plt — Top Row — Toto, Price, Satterswaithe, Hill, Wales, Keatts, Gooding, Krause, Middle Row — Serrano, Coleman, Bottom Row — McInvale, Allen, Miller

Top Row — C Co commanding officers; missing are CPT's Bill Parish and Melvin Morrow; Bottom Row — 3rd plt leaders and PSG SFC John Partee

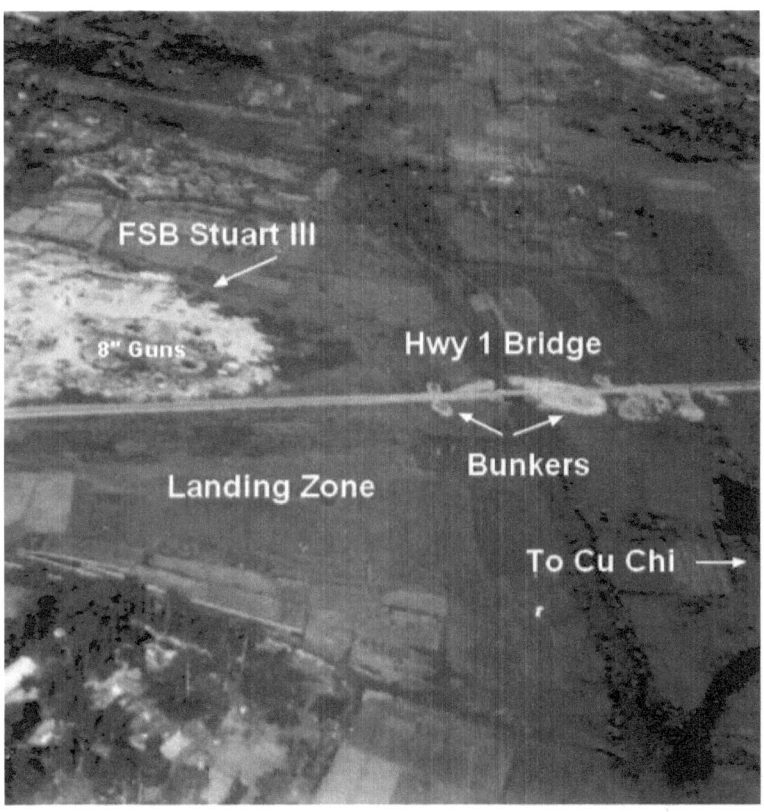

FSB Stuart III outside of Trang Bang on QL-1 or Hwy 1 and bridge bunkers

FSB Pershing, north of Trang Bang. Home of 2/12th battalion from Oct '68 to Feb '70

SGT John Spoores and Barry Price in Dau Tieng, April '68

YOUNG MEN IN HARM'S WAY | 167

Barry Price, Arturo Quintanilla and Ron Stepsie (KIA, Dec 14, 1968)

Russell Zimmerman, George Toto, Pat Flood, Terry Corum,
Richard Coleson at stand down, in Cu Chi

Part of 3rd squad at PB Granite, Feb '69, Back—Russ Zimmerman, Curtis Tassen, Clarence Carson, Joe Starling in front

*3rd plt—Back Row, Coleson, Mikita, McInvale, Potts,
Front Row, Gries, Tassen, Bowden*

2nd squad, John Mikita, Vernon Becker, Darrel Kuhnau, Bruce Solem at FSB Pershing, Jan '69

3rd squad, Collins, Tassen, Starling, in back, McInvale

YOUNG MEN IN HARM'S WAY

McInvale, Mikita, Bob "Doc" Meyer, Bowden, sitting is Jochum

Standing, Price, Wales, Buckley, Hill, sitting is Allen, Krause, Toto

Eagle Flight with 116th AHC "Yellow Jackets" at FSB Pershing—Dec '68

13. LIFE IN A FSB

The location we were dropped into when the Battalion was moved from Dau Tieng down to the southern area of Hoc Mon was comprised of lots of rice fields and small villages or hamlets. Our laager site was also a fire support base. We were sharing our location with our heavy weapons platoon and their 81mm and 4.2 inch mortars. To our west was a small village located only a few hundred meters from our perimeter and right next to us was a cemetery. This area was only five miles from the city limits of Saigon, so urban sprawl was prevalent and population density was heavy.

In the center of our laager was a huge deciduous tree which I estimated to be 80 feet high or more. Someone had taken it upon himself to climb up this tree into the canopy and fasten a flag pole that jutted out above the top of the tree with an American flag attached to it. It was an impressive sight to see.

The perimeter of the FSB was sprinkled with firing positions made from sandbags which provided protection from incoming mortar rounds and small arms fire. If the site was to be used for a long period of time, PSP or perforated steel plate was brought in. That was used to construct temporary runways for airplanes, but for us, it was used as a reinforced roof with three layers of sandbags placed on top, each bag pounded flat by a trenching tool or shovel.

In the center of the perimeter was the company or BN HQ bunker, a FDC bunker for fire control, and possibly a commo bunker for radio communications. Sometimes there might be a mess tent too. Our battalion commander, LTC Don Green flew in to check on how operations were going. He spent some time chatting with 1LT Jimmy Ford, our X.O who stood there shirtless, arms folded, listening intently. Ron Hendricks had resumed his duties as our C.O. since his recovery from wounds on June 9th. Jimmy was from a small town in Georgia. He had plenty of combat experience, so he had the Colonel's ear. I wondered what was being discussed. For those of us void of "rank", having the brass drop in on occasion did not pique our interest.

To improve our security, the engineers brought in a bulldozer and created a berm around our laager site. That would give us some protection to move around and be concealed from light arms fire should we be attacked. Right after they completed this task, the rains turned the road into a quagmire and the dozer was stuck in the mud up to the tracks and had to be winched out. The good news was we had a dirt shield to protect ourselves. The bad news, especially now that it was the monsoon season, this same berm now transformed our laager site into a lake with the accumulation of water from the daily rains. We were literally living in a lake. We used our ponchos to create some shelter from the rains. Steel fence stakes were pounded into the ground with one tied between the two so we could cover those with our ponchos to create a tent shelter and we built beds and suspended hammocks we had purchased from the locals to sleep in. The guys spent time over at the mortar platoon where we picked up empty mortar boxes. We used these to build a platform to stand or sit on so we could get up out of the water and dry our feet.

Our health and hygiene started to become a problem for those of us who had been there awhile, not so much for the new guys who had just arrived. Soon, they would find themselves in the same boat. Initially, when we were operating out of Dau Tieng, we were close to clean clothes, showers, dentists, and doctors, but not now.

For me and others who arrived together, having been here for, well, going on five months, we adjusted to drinking the water, not local water, but our treated water that we had access to which was airlifted to us in water trailers. No one really tested the native water supplies unless under duress and then only if we could put halcyon or iodine tablets in the canteen to treat the water, then it tastes like dirty underwear. The water wreaked havoc on the digestive tract, giving you dysentery, but then, so did the hookworms you picked up from wading in the rice paddies and water buffalo manure.

The landings from helicopters into swamp areas exposed us to the snakes and mostly leeches that found a way into our clothing in spite of tucking our fatigues into our boots. When we got out of the water, we inspected every inch, private areas included, for passengers, then removed them with a lit cigarette butt or a squirt of bug juice as we call it or insect repellant. Occasionally someone would get some type of bug or insect bite that was painful as well.

We didn't shower with all the rain in the wet season, and when we did get a chance, it was more like a sponge bath. It was a great feeling to wash all the dirt and sweat collected over a few weeks, and then to be able to pull on a fresh set of socks and fatigues. Haircuts were done by someone in the platoon who either had some experience in the trade or was just willing to give it a try. Short hair was easier to deal with under stressful situations when sweat was pouring off the brow as well as washing it to keep any critters out of it.

Personally, I should have brushed my teeth more often, provided I could keep track of my toothbrush and toothpaste. I had to see the dentist in July and fixed a few problems. The doc said to me to take care of those teeth, like I had access to that stuff every night. We shaved when we wanted to. For most of us, shaving was not needed because many had baby faces, me included. Still, there was an occasional whisker or two that needed trimming. Living out in the bush was equal to the early settlers, I imagine, when they crossed the plains of America. You made do with what you had, which was little. The medics tried hard to keep us taking our medications. There was nothing they could do for the weight loss each of us experienced. No amount of calorie intake can keep up with what we were burning off every day and the Army was not known for its cuisine cooking nor would we find it in a box of C-rations.

There was one exception to this weight loss that everyone experienced. Years later, Craig Schoonderwoerd would claim that Billy Zimmerman was the only guy he knew that remained fat. I use the term fat only to imply that he still looked stocky and like he hadn't missed a meal. Craig claims this was due to the fact that Billy loved to eat ham and lima beans and while everyone else was avoiding that particular box of C-rations, Billy would eat anyone else's supply that they wanted to give away.

One afternoon, I started to head over to the other side of the perimeter for something when it started to rain, and rain it did. So heavy was the rainfall that it stung the skin. By the time it stopped raining, and it seemed like less than an hour, the water level in the laager site rose about six inches in depth. I had started out with the water at ankle depth. By the time this downpour ended, the water was above my calves. This only added to our existing misery and that was staying as dry as we could while trying to rest and sleep and maintain some kind of normal life. The next morning, Steve Ward was standing in knee deep water adjusting his mirror he had hung on a pole as he attempted to shave. His blond hair was a striking contrast to the early morning sunlight filtering through the large tree that dominated the center of the laager site.

Fire missions were scarce for the heavy weapons boys. We found very little enemy action initially. The thought was the VC pulled back from the TET Offensive, having been severely defeated at every location he struck. Losses were estimated at around 55-60,000 men for the VC. He had to withdraw to regroup, resupply, then plot his next move. That was why we were brought down to this area to act as a line of defense by putting up a picket line all across the approaches from the north off of the Ho Chi Minh trail down toward the Capital. This was the main objective for the North, to defeat and take control of the city of Saigon in battle. If the city fell, so did the government of South Vietnam.

Someone at Battalion in S-3 had come up with a new twist that was shared with the company. We were going to go on roving ambush patrols as a platoon and set up three ambush sites per night. You could hear the bitching for miles. This new wrinkle started that night, fortunately for our platoon, we were not in the rotation to get AP today.

Not only were we operating out of our basecamps, but we are setting up PB's (patrol bases) as well. From these PB locations, we occasionally spend the night then swept the immediate area before we moved back to our basecamps. Or, we might also run night operations from them as well. The PB was just a place to stay temporarily. It was not defended nor were any forces stationed here. We just came and went as needed.

We were not in this area even a week and the night patrols began. The point man started us out on the mission. He was being directed by his squad leader who had the azimuth (compass direction) heading for the first check point on the map. The going was slow and hard. The ground had been plowed by the local farmers. It was that time of the season when the rice paddies were being prepared for planting. Each step was an effort to extract your boot from the gooey mud only to return it to the ground again and again. It was difficult to know how far we had traveled. To get to our check points meant knowing how many meters we had traveled. We had no reference points to go by. It was what made night navigation difficult because you cannot rely on geographical shapes of terrain, but needed to figure out where you were using roads and trails and sometimes village lights. We guessed that we have arrived at our first AP site and I radioed our location into the battalion.

Night time brought certain protocols into play for each element of a battalion, company, and platoon. No matter where you were, you were in constant contact with the chain of command. At night, whether we were in a FSB, or out on ambush patrol, we were required to provide a SIT REP, situation report, every fifteen minutes once the *watch* started or when the AP had reached its destination and had set up the ambush. Out on an AP, remaining quiet to avoid detection was paramount, so there was little if any talking that took place. Every AP or LP carried a radio with them. When you were contacted on patrol for a SIT REP, you would hear, "Charlie 36 X-ray, Sit Rep." That would be repeated if there was no response, i.e. someone fell asleep when they had the *watch*. This was not good. If the SIT REP was negative, nothing to report, the response was to press the TRANSMIT KEY on the horn twice, meaning you were breaking squelch twice. It sounded like "*SISS*" "*SISS*" on the other end. You would get a "Roger, Out" when heard. Everyone pulled watch, a two hour period when your eyes were glued to the field of fire to your front. Being on watch meant you were responsible to stay awake and alert and WATCHING for enemy movement. The lives of your fellow soldiers was dependent on you. Keeping guys awake was difficult at time, depending on how much rest we were getting. Falling asleep put everyone in danger. Watch was done inside a FSB or laager site too. Every platoon had someone awake at all times at night. That held true for every squad within the site as well. The primary reason we were so tired was the fact no one got a full nights rest unless you had the first watch which started about 2200 hrs and the last watch which started at 0400. Of course, the more men you had in a squad, the easier this routine became.

An hour later, we broke up our AP and moved to our second location and repeated the process. Both moves were to the west and now we needed to execute a 180 degree turn and head for checkpoint 3. We walked and walked in the sticky mud. Keeping pace count was a difficult task. We even had two guys keep count so they could compare numbers.

Our PL, R.W. "Bud" McDaniel, made a decision we had gone far enough and looked for a likely place to set up our third ambush. We were supposed to be accurate and KNOW where we were. How else could we call for artillery support when it was needed? Did we want it raining down on us if we had made a mistake? The Lieutenant was sweating this one out. After two hours, we pulled out and trudged on to our next check point. Again, we followed orders and waited it out. Time was dragging by and we were all uncomfortable because it was so light behind us from the flares being dropped around Saigon. Perhaps they had some enemy probing going on that day. It was around 0200 hrs and time to move on to our next check point. As dawn approached and the morning sky begin to light up the countryside so we had clear visibility, without any activity, we broke down our ambush site and trudged back to base for some hot coffee and some shuteye. In a couple of days, we would be repeating the same thing. But tomorrow, it would be another platoon's turn for night patrol.

Our supply chopper arrived, bringing in fresh water and some basic supplies. Included with hardware was mail from home and our goody pack of treats. About an hour later, the platoon was gathered together for mail call. Not everyone got a letter or something from home, but we all stood around eagerly waiting for our names to be called. I got a letter from Mom and Dad, and one from Pam. She was the sister of Judy, who I was dating before I went into the service. Shortly after I entered boot camp, she sent me a *dear john* letter, remember me mentioning that? So much for love, I guess. Anyway, Pam, her sister, started writing to me for some unknown reason. She always wrote an interesting and informative letter. She probably wrote more letters to me than anyone else. It was very sweet of her to have done that. Pam stopped writing to me once she knew I had made it back home safe. I don't remember ever seeing her after I got back, but I will not forget her caring support while I was in this country.

We spent the day getting caught up on correspondence as I ground out a few letters to my parents, Pam, and one to Patty Pressley, the gal I was hanging out with when I was home on leave prior to leaving for Nam. I met Terry Corum, from Huntsville, Alabama who arrived as a new NCO. He was slim built when you saw him with his shirt off. He had dark curly hair and had a heavy southern accent when he spoke. He got a quick ribbing over that.

Down time is important. It was a chance to relax and drop your guard for a while. We discarded our fatigue shirts to let them air out while we pulled off our shoes to dry out our feet and look them over for signs of infections. A few guys had sandals or flip flogs they had purchased from the villages to wear. They rolled up their trousers if they were going to drag on the ground.

Someone would wander by from another squad or your bunker buddy would start up some idle chit chat, and some basic jabbing. SGT Bishop, Serrano, Barnes and Gooding were going at it, goofing around. Beltran was busy writing home to his wife. He was a bit reserved and it was hard to see him as a biker. He lived near me in Fremont, California, in the Bay area. Bishop was hanging out playing cards, an activity he was usually involved in with Brasher, that and downing a can of Miller beer. Percy Miller, Lynwood Keatts, and Steve Ward were discussing life at home. They were just getting accustomed to being in the platoon. Hill and Price stopped by for a bunker visit. We tried to keep our focus on the happier times as we each told stories about ourselves. Everyone eventually took a turn at getting some shuteye. We were not scheduled for patrol today since we were out all night. The company was on standby as a reactionary force in case some other unit got in a bind.

As evening fell, we listened to the radio and continued with the small talk. One of my favorite songs came on and I turned up the volume a little bit as I heard the first line of the song, *I'm a girl watcher, I'm a girl watcher, watchin girls go by, hey my, my.* The squad was gathered around a bunker and I was sitting on the ground. I found a dry area and I was resting as I propped myself up against the sandbag wall. I felt something crawl up my pant leg and suddenly I felt a sharp pain or sting. I quickly jumped up and was frantically slapping my pant leg. I didn't know what had happened but there was a searing pain on the inside of my right thigh. I was in a lot of pain. My leg felt like it was burning. The pain subsided after a few hours but I was still in a lot of discomfort. In the morning I found a large centipede near the place where I was laying. My leg had a large red welt where I was bitten. I felt better but it would be a day or two before I lost the sensation.

Charlie used the darkness of night for most of his activities. His troop movements and resupplies were done in the cover of the night. This action could only be checked by massive night operations. Mostly this was made up of ambushes set up along routes hoping to find the VC using an ox cart trail, small path, or road near a village. Our intelligence gathering efforts dictated where this was done. Because of the monsoon season, enemy troop movement and contact was running at a minimum. The VC seemed to be avoiding engaging our forces.

About the only newsworthy events taking place in July, were the occasional munitions and food cache discoveries made by patrolling ground units. The area we were in had too much standing water for any mechanized units. They cannot venture off the roads and MSRs without getting mired down in the mud. It would be this way until October. Any additional firepower we may need at the point of attack would have to come from our artillery or air cover.

Today was July 14th, and we were west of our night laager site about 5,000 meters. After patrolling the area our orders were to set up at the patrol base at XS710032. Later we got the word from the first sergeant, SFC Kaplan that the company would be pulling AP's tonight. McDaniel gave the squad leaders a briefing of the nights' AP. The operation called for three roving patrols made up of one platoon each. AP1 would start out to the east then execute a 180 degree turn and work back toward the PB. AP2 and AP3 would start west, with AP2 moving further west than AP3. AP3 after moving west to its first two checkpoints would reverse its direction and travel east for its next two ambush sites. AP2 and AP3 would each execute a cloverleaf movement that returned them to our PB and beyond. It looked like a very screwy maneuver fraught with potential disaster. We abandoned the PB around dusk, around 1730 hrs and checked our maps and compass. We were to start at dark.

Some of the guys took time to slap on some *bug juice*, our lingo for insect repellant. It was unpleasant and oily, but was effective for mosquitos and did a great job of removing leeches when they attached to you. Just a small squirt and they dropped off. Everyone saddled up, and I made sure the radio was checked, and the word to head out was given. I was trailing McDaniel and the pace was slow. 1st squad had the point with SGT Hugh Bishop keeping pace count. Thirty minutes into this patrol and we arrived at our first AP site and quickly set up our ambush. Two hours later, we broke down and moved on to our second AP; so far so good.

Around 2100 hrs, we left for our third AP. We walked and walked and we seemed to be off our mark. Trying to keep an accurate accounting of the distance we had traveled was difficult and compounded by the heavy mud and water of the rice paddies we were crossing. It seemed the patrol was unsure of exactly where we were. We pulled up and tried to find some reference point in the dark but couldn't make out any distinguishing landmarks.

SFC John Partee who had just taken over as our platoon sergeant, whispered to McDaniel, "Let's hold up here for a second," he said. "Send SGT Bishop back to see me." A minute later Bishop showed up. Partee asked him, "What's our pace count?"

Bishop replies, "I'm not exactly sure. It's hard keeping an accurate pace in this muck. I'm pretty sure we're at or about the point where we need to turn on our new azimuth."

"Okay, then let's do it," Partee said and McDaniel agreed.

Bishop returned to the front where his point man was and the platoon made a 45° turn. As we moved out on this new azimuth, there appeared to be a tree line we could make out in the starlight. Behind us was the glare of Saigon's sky, lit up by aerial flares. We were a hundred meters from the trees and we trudged forward, each soldier stepping in the footsteps of the guy to his front. The column was in single file and we had no flankers out.

As we approached dry land, we were maybe 20 meters away. *KABOON, KABOOM, KABOOM*, in rapid succession, three Claymore mines were detonated. *BAM, BAM, BAM*, and then our platoon was suddenly receiving automatic weapons fire. *Pap. Pap. Pap, Paaaaaaaaaap, Pap, Pap...Paaaaaaaaaaaap. Paaaaaaaaaaap, Paaaaap* from the machine gun and the loud Taaaaa dut, Taaaa dut Ta dut,.... Taaaaaa dut, Ta dut, the loud crack of rifle fire. Everyone dived into the water.

"What the fu**," Carlos Serrrano screamed.

Then I heard in rapid succession, "Return fire, everybody fall back, stay low, shoot the SOB's," yelled McDaniel.

We were too close to the kill zone and had no cover. We tried to pull back. We were cussing and yelling to each other to pull back. I was wrestling with my radio and trying to keep in contact with McDaniel and Partee. The guys forward of me were still screaming and Quintanilla is cussing up a storm. I could hear the bullets whizzing by my head like mad bees. I was bent over trying to keep a low profile while I attempted to back pedal through the rice paddy.

In all this crazy frenzy, we could hear other English voices. They were not ours. We quickly figured out we had walked into one of our own AP's. It seemed like it had been minutes but may have been only a matter of seconds through all the shooting and screaming that we started yelling.

"Cease-fire, cease-fire, cease-fire, we're Americans!"

The shooting stopped and all was quiet. Everyone slowly rose up off the rice paddy.

The guys from the ambush yelled out to see if anyone was hurt. Both groups were dazed from what just happened. McDaniel finds out it's an AP from A Co that fired on us.

"Anybody hit?" Partee yelled out. "Is everybody okay? Let's take a headcount."

Arturo Quintanilla and Steve Ward, along with Craig Coleson, were standing next to Keatts watching Partee milling about. It was hard to see faces in the dark.

"OK, sound off 1st squad," Bishop called out, then heard replies from Dixon, Coleman, Christianson, Kuhnau, Quintanilla, and Ward. This was followed by Conlin's squad as Allen, Miller, Rinkle, Keatts, Polus, Hill all sounded off they're OK, then it was Freidig's squad's turn.

"Where's Freidig and the new guy?" SGT Bishop asked, "Anyone know..." but his sentence was stopped short.

"Two down over here," Richard Polus yelled out. It was Freidig and McGeath, the new guy. Clarence Olson, our medic, hurried over to where Friedig and McGeath were laying, face up. We put a light on McGeath so Olson could check him out for wounds, but he had no pulse. There was a bullet hole in his neck.

"He's gone," Olson said, then turning his attention to Friedig, asked him where he's hit? Freidig pointed to the hole in his chest leaking blood, which Olson applied a dressing to.

I keyed the mike on my radio and called Charlie 6 X-ray and gave the RTO a sitrep of 1 KIA and 4 WIA and requested a Dustoff at our location.

Not knowing exactly where we were, we had to talk the chopper to our location by listening for his rotor blades and coaxing the pilot to change directions toward where we were. To aid the pilot in finding us, we put a strobe light in a steel pot and held it overhead.

"Charlie 36 X-ray, this is Dustoff 27, over."

"Dustoff 27, bear left of your location. We hear your approach, over."

"Roger Charlie 36 X-ray, standby."

"Dustoff 27, this is Charlie 36 X-ray, on course, will provide strobe signal for bearing, over."

"Roger Charlie 36 X-ray, have you in sight. Is LZ hot?, over."

"Negative Dustoff 27, LZ cold, we've got one kilo and two whiskey's, over."

"Dustoff 27, Copy that."

After William Potts, Robert Beltran and Russ Zimmerman load McGeath with the wounded into the chopper and the Dustoff departed for Cu Chi and 12th Evac, the radio crackled with incoming traffic.

"Charlie 36 X-ray, this is Charlie 6 X-ray, need a sitrep and Charlie 6 says to hold there after the Dustoff until daylight then return to base, over."

"Roger Charlie 6 X-ray, copy, evac complete, Charlie 36 X-ray, out."

McGeath had taken what appeared to be a round in the throat. I think he was hit with a Claymore. He was a new replacement and this was his first operation. His death was listed as a *misadventure* in the official records. That's how the Army listed those killed by friendly fire. How was the family going to understand what had happened to their son when they got the word of his death? He just got married before he came overseas. After arriving at 12th Evac Freidig would later be transferred to Japan where he would spend two months recouping from his wounds before returning to the company.

There were two others wounded, but nothing serious. The whip antenna on my radio had been severed in the melee. That antenna was only a few inches to the right of my head. I said a prayer for Richard McGeath and for myself as we stood around and gathered our thoughts along with the guys for A Co.

"We could see you moving across to our front from a long ways off," one of the guys from Alpha Company said.

"We were on our way to our own ambush site when we spotted the movement. We would never have seen you if it wasn't for the flares over Saigon," he said. "Our SGT gave us the orders to hold up where we were and to set up an ambush and then see what happened. After that the rest was just history."

We didn't let them down. I don't know what the fallout was from this event. I don't remember getting interviewed after we returned to our NL to discuss this unfortunate event and what to do to avoid something like this happening again. It was just a part of war and shit like that happens. We found out later that we overshot our third AP site by almost 2500 meters. It was really fortunate that we were discovered by this patrol from Alpha Company. If we had turned left when we had overshot our mark by 2000 meters, we would have walked directly into Alpha and Bravo's basecamp. The next day SGT Terry Corum is assigned to take over 3rd squad in place of Dale Freidig.

Not too soon after that, I wasn't feeling so good and went to see the medic. I was running a fever and my glands were swollen, including my testicles. One of them was the size of a golf ball and with no underwear this thing was swinging around like an anvil and giving me a lot of pain. I was sent into Saigon to Bien Hoa to a hospital where they gave me some antibiotics and they kept me under observation for a few days until my hardware returned to normal. Upon my release, another guy whom I cannot recall his name from C Co and I wandered over to Bien Hoa AFB to see how they lived. There were paved streets with curbs and sidewalks and they had their own barracks. They had a movie theater and across the street there was a snack bar that served hot food, hamburgers, and drinks. We opted to see what they were serving and got a burger. It was not like McDonalds or Wendy's, but it was hot and fresh. We could hear two Air Force enlisted men sitting at a table not far from us complaining about having to share their quarters with someone else. The bantering continued from the boys in blue sitting around us. One guy went so far as to say to the other airman, he had to listen to music on his roommate's boom box and he played it too loud.

We laughed at listening to how tough they had it. Why on

occasion, they even got a stray 122mm rocket that hit the base. We reflected back on our own situation and could only swear in silence at these guys. They didn't really know how blessed they were. We finished up with our chow and checked out. In the morning I was back in the field with my unit.

On my way back to our laager site, I got a chance to replace my wrist watch. I found a vendor and picked out a Seiko watch with a date calendar. Who knew what kind of movement this watch had on the inside. I had heard they can do anything just to get them running. I took my chance with my selection and gave the man $5 in Dong currency. It was a nice looking watch, gold in color with an expandable wrist band. It should do.

July 27th and 1st platoon had AP duty. Working at night was not pleasant. The rice paddies were flooded and difficult to traverse in the daytime, let alone at night. I was not sure who had the platoon, either 1LT's Pat Kiggins, Robert Norris, or Laurene Johnson. Platoon leaders were arriving and departing way too quickly for any kind of leadership or combat experience. As it turned out, SGT Michael Granum would handle the ambush.

In a rice paddy west of our night laager, PFC Lawrence McCloud experienced a moment of truth he will never forget. McCloud, of Greenville, Miss., was one of 15 infantrymen from 1st Plt. on a night ambush. Shortly after dark, patrol leader Sergeant Michael J. Granum of Portland, Ore., reported silhouettes moving on dikes about 150 meters from his platoon's position. Soon the G.I.'s began receiving small arms fire. From behind a two foot embankment, the Warriors watched as five dark shapes rose against the night sky.

McCloud saw three of the enemy converging on his position. "One of them was carrying a loaded RPG rocket grenade launcher on his shoulder. I was fixin' to shoot, but I could see that RPG pointing right through me," he said.

McCloud waited anxiously until the three V.C. were beside his position, only 15 feet away, and opened up. The figure with the rocket launcher whirled and squeezed the trigger.

"I thought that was it," recounted McCloud, "but nothing happened."

Silence followed as the G.I.'s held their ground. They heard movement periodically during the night, but couldn't see anything. At sun-up, McCloud ventured from his slippery position behind the dike and found the anti-tank weapon, still loaded, laying half buried in the mud. A dent in the primer of its propellant charge indicated that the round had been a dud. No one was injured in the action, and no enemy dead were found.

The entire month of July was spent sweeping through local villages checking for VC infiltrators, running traffic check points, and sweeping the surrounding area during the daytime, and running night patrols. There was almost zero contact with the enemy. They had seemed to have vanished from the area. When we were on night patrol, we returned to the basecamp and caught a few winks, then around noon, we conducted a RIF of the area, returning after several hours to await orders for our next AP assignment. We never even climbed into a Huey. The rest of the company got a few rides, but for some reason, our number was never called. Maybe our opportunity would change in August.

1LT Richard Wiggins who would soon become commander of Alpha Company had this to say about our stay near Saigon:

> *We did have an assignment close to Saigon in the area they referred to as the rocket belt and we used to run a lot of patrols from our bases. We didn't have any artillery in the base we were on and relied on support from FSB's in the area. I can remember on numerous occasions we would receive fire from the VC but we were*

restricted from returning fire with anything larger that our individual weapons until we received approval from the U.S. Ambassador, which would take hours and by the time we received permission the threat was long gone. Dumb, dumb, dumb and I remember how all of us cussed that requirement. We did a lot of night patrols while there and on most of them we would move every couple of hours which was pretty hairy because we had a whole bunch of platoon sized patrols that were moving at the same time, something none of us liked doing in the dark of night, especially since night time navigation wasn't exactly an accurate maneuver. I remember vividly arriving at our ambush site one night and we were in water up to our waists cussing the S3 folks that picked that location for an ambush site. I kept checking my watch to be sure when the time arrived to move we were moving. On several of those time checks I had leaches on the face of my watch.

July ushered in more replacements: Robert Beltran, from Fremont, California, Terry Corum, Woodville, Alabama, a shake and bake E5, Richard McGeath, Murphysburo, Illinois, Arturo Quintanilla, San Antonio, Texas and Russell Zimmerman, Charlotte, Michigan joined third platoon. A total of 23 were added to the company. Willard Brasher, who was one of the RTO's, headed for home. He was a character, thin as a rail, with dark glasses and had a good sense of humor.

14. MEDCAPS, TRAFFIC CONTROL, CORDON AND SEARCHES AND AP'S

August arrived but nothing really changed. Charlie remained a low profile figure who only on occasion took a pop shot at us or was caught in small numbers moving from village to village.

The rains were heavy at times and I felt like I was living in a swamp. Water was everywhere and nobody could dry out. We took every opportunity to keep our shoes off and our feet out of the water to avoid any kind of skin rot. One day it rained so hard that it stung the skin. It was almost impossible to even stand in the rain. I swear after about an hour or so, the water rose in our basecamp eighteen inches. It might have been good for fishing, but not for camp life. Yep, life was good in the infantry. Except for ambush patrols at night and our routine RIF's around the laager site, it was quiet for the first ten days of August.

Scuttlebutt was floating around the laager site that there was a change at Division. MG Ellis "Butch" Williamson was assuming command of the 25th ID, taking over for MG F.K. Mearns. Commands like these don't change often, as they are usually a one year assignment. I'm sure there was a big ceremony with companies in clean uniforms, flags furling in the air, and the band filling the air with some military march. There is a lot of pageantry to these events.

One of our main missions in this area was to conduct cordon and search operations on villages in the area. We must have pulled off that maneuver a half dozen times over the past month trying to catch the VC either using the village as cover or maybe visiting relatives. In any event, we would seal off a village in the middle of the night, usually around 0300 or 0400 hrs and block all exit routes and then wait till daylight to sweep through the village searching every hut and tunnel. We would do this jointly with the local ARVN National Police forces or local Popular Forces who would conduct the interviews or interrogations. This would usually take most of the day before we would be released and return to our NL. On a few occasions we would catch suspects fleeing the area and take them down.

Other policing activities which we conducted were traffic check points, where we validated ID's and inspected vehicles for contraband. These were set up at various road locations and lasted most of the day. It was considered to be low key. The VC were avoiding us most of the time and refused to clash with our forces in the area. So far, there had been no mounted effort to penetrate our defenses to gain access to the Saigon area. Because of the lack of enemy activity, the brigade moved forces further from our line of defense during the daytime, conducting RIF's in areas normally left alone. That required the utilization of aircraft and some of the companies were involved in more airlifts.

I had barely broken in my SP4 patches on my fatigues and was promoted to acting SGT E5 effective August 2nd. In order for this to happen, I was sent back to Cu Chi to Leadership Development School, to begin on August 9th. I was supposed to go back to Long Binh and catch a flight up to Cu Chi on a de Havilland C-7A Caribou. Right after we took off, the plane developed engine trouble so we turned around and headed to Vung Tau where the aircraft are serviced. Apparently that's the only spot where these aircraft had maintenance. We were only there for a short time when we climbed aboard a new aircraft and took off and headed for Cu Chi. Once I got off the airplane, I walked over to where the combat leadership school was and reported in. I got instructions where to stay and my training started the next morning. My instructor was an E6. They tried to teach us everything we would need to be good leaders. The school lasted for nine days.

1. Map Reading 8 hrs.
2. Forward Observer Procedure 6 hrs.
3. Night Firing Techniques 4 hrs.
4. Booby Traps 3 hrs.
5. Ambushing Techniques 2 hrs.
6. Use of gunships 2 hrs.
7. Demolition's 3 hrs.
8. First Aid 2 hrs.
9. Communications 2 hrs.
10. Combat Intelligence 1 hrs.
11. Leadership 2 hrs.

While I was off at leadership school, the company finally got some air time. The morning of August 17th the company made a combat assault, reporting zero results and on August 18th and 19th they CA'd along with Alpha Co to sweep an area to the north of Hwy 1, then return to our NL.

Upon graduating from the leadership course, back to the field I went. I gave up my RTO duties and assumed the command of 3rd rifle squad leader on August 21st. I needed to get to know the men who were in my squad. I had George Toto, Russell Zimmerman, Tom Sovey, along with Robert Beltran, and Ed Wales. I quickly learned that I needed to keep an eye on Sovey. He was not the sharpest guy and seemed oblivious of his surroundings and the dangers he faced. On this day, the 1/27th was pulled out of the area and sent back to Dau Tieng, leaving the 2/22nd and the 4/9th supporting our right flanks.

My first mission as a squad leader was August 22nd and it was to take the squad out on an ambush patrol with the platoon, then around 0400 we moved to cordon off a village for a search in the morning, along with Alpha and Delta Companies. Over the next two days, August 23rd and 24th we conducted night operations using ambush patrols and then after returning to our NL and resting up we would conduct small sweeps of the area in the afternoon. August 25th the 1/27th returned from Dau Tieng. On the morning of August 26th, we left our night laager at 0500 in order to move into a blocking position at XT705995 and arrived and were set up by 0730 hrs. We spent the whole day there waiting for something to happen, but nothing did, and we returned to our night laager. What was supposed to happen when you were a blocking force was to have another element sweep the area to your front, hoping to drive the fleeing enemy to your position where you could then engage them. That was the operational plan anyway.

August 29th we left the NL at 0830 to conduct cordon and search via a combat assault with Alpha Co near Xom Giong Sen (1) out in the marsh area west of our line of defense. It was a suspected VC sympathizer and was located near low land marshes. These marshes separated the Hoc Mon area to the east and Duc Hoa which sat on the far side of this sizable geographical wet land. Normally we did not roam this far west, but today was today and we were sweeping in with our eagle flight barely 50 feet above the rice paddies to quiet our approach on the village. I was sitting in the doorway with my feet dangling down into the breeze. It was a thrill ride today and the pilot was having his jollies. The company needed a surprise arrival so we could trap any VC that may be hanging around the village. By coming in low, the noise of the choppers wouldn't be detected so quickly. The sound of the rotors striking the air was always a familiar sound to friend and foe alike.

All was quiet as we approached the LZ and deployed from the Hueys. A quick sweep of the village caught one VC attempting to dart out the back side. He was stopped with a quick toss of a grenade from a G.I. in 2nd platoon. As we looked for weapons, we checked everyone's official ID cards by asking them for their *can cuoc* (con-cook). The search was complete so we moved to the checkpoint beyond the villages and straw grass hooches. I had to remind the new guys not to bunch up. That was just inviting trouble because one RPG or burst from an AW would spell multiple casualties.

"Toto, you and Zoom (Zimmerman) get some distance between you. I don't want to repeat myself again, and when you're the rear guard, keep your eyes on your job and what's behind us and let me know if you see anything suspicious going on. Got it?"

"OK, Sarge, I copy," Toto quickly replied.

"We'll see," I muttered to myself. I know they're new and will eventually get it, I just needed to make sure it doesn't cost them their lives in the meantime.

Being that it was midmorning, we could see the haze of blue smoke that curled upwards from the cooking fires. Most of the stares back from the locals did not appear to be friendly and the women and children remained aloof as we left this village behind us.

The radio crackled and SFC Kaplan told us we were to search the area along the edges of the swamp and hedgerows for food caches or anything else the VC may be hiding. The company fanned out and we started to search the vegetation along the banks. SGT Keatts, Pat Flood and John Mikita were searching a promising area and it was not long before they discovered a small cache of 82mm rounds. A bit later, SGT Conlin and his squad were working the other side of the waterway when Miller and Becker found a few more 82mm rounds. Rather than load this stuff up on a chopper, McDaniel radioed to our C.O. Hendricks what we found and Hendricks decides to detonate what we had found on site. It was safe since we were now some distance from the village.

A few chunks of C4, a primer cap, and a foot of fuse cord was all that was needed to get the job done. The company was pulled back to a safe distance, and "fire in the hole" was sounded. Thirty seconds later, the charge went off and the munitions were destroyed. We searched for several more hours with no luck. 3rd platoon had point duties. McDaniel told me my squad had point and gave me the azimuth so I could check my topographical map and identify land features to verify we were on course. "Wally (Ed Wales), you're on point. I've got your back. Let's head out when I give the nod toward that distant tree line," as I pointed the direction I wanted him to take us on. I kept him on a steady course and he led us to our pickup point. Today, we caught a ride back to our NL. Our taxi was the 116th AHC *Yellow Jackets*.

The next morning, the 29th we left our NL at 0254 with Alpha Co and the 53rd RF (regional force) and moved out to a little village to our west, Quang Trung. We spent most of the day searching the village and immediate area but discovered nothing. We duplicated the same action the next two days with other small villages in the area with the same results. At least the guys in my squad were gaining some experience and weren't under fire while they learned about communications, spacing, and reading the signs and people around them.

After learning how to use the flamethrower back in April, my second training class had me learning about how to handle explosives. I was sent to demolition school for a day after I made sergeant in Cu Chi. The training involved using the various plastic explosives, blasting caps, fuses, and detonation devices used by the engineers. They were not always available where we needed these skill sets, so having someone in the company who could handle these tools was important. It was expedient that we destroyed munitions as we discovered them. With so much unexploded rounds lying about the countryside, it was a tempting device for the gooks to use as booby traps.

There were two types of plastic explosive, Composition C-4, a white material with the consistency of clay containing 91% RDX, Research Department Explosive (nitroamine), that was in the shape of a bar, 2"x11"x2 1/2" weighing 1.25 lbs. It could only be set off by a combination of extreme heat and shock. The second plastic was green in color and came in flat sheets about 3" wide and 11" long and was thin, roughly 3/8" thick. It was not as pliable as C4 and I don't know what it was called. It was rarely used and I only remember being trained with it.

We learned to attach blasting caps to various types of fuse cord which had different burn rates. The last element was detonation cord or det cord, a white plastic cord used to link several charges together so they would explode at the same time. It was a good class that served me well in the future.

Replacements who arrived in August: Dennis Buckley from Burbank, California, Richard Deimler from Hummelstown, Pennsylvania, Pat Flood, New York, Primus Gooding, Jim McInvale, Alabama, Henry McKenzie, Tom Sovey, Seattle, Washington, George Toto, Pennsylvania, a total of 25 were new to the company. The turnover ratio due to wounds, fatalities, or rotation back home took its toll on experienced leaders in the field.

15. SOUTH VIETNAM'S MILITARY UNITS

Besides the regular army in the South which was the Army of the Republic of Vietnam or ARVN for short, there was also the ARVN Rangers, which were a very dependable group of soldiers. Most of the combined operations the Battalion or company were involved with were with ARVN Ranger units, organized and ready to fight. Less dependable were the regular ARVN forces we saw scattered around the countryside that were stationed at various ARVN compounds which were located within and outside of towns or main junctions on roads.

These compounds housed small unit forces, platoon size or less. Whenever we saw soldiers who were assigned to these posts, they seemed rather indifferent to the situation their country was in. They wore flip flops, no helmets, and usually had their weapons slung over their soldiers and only one ammo clip. The VC usually left these guys alone like they had a secret pact with them. Popular thought with the troops was that they had a "I won't bother you if you don't bother me," pact with the enemy.

The next group or tier of support elements had four types. The first was the South Vietnam Regional Forces, originally the Civil Guard which were a form of local militia put in place to counter the local VC forces. They were recruited locally and were more aligned to the provinces they operated in. They received 13 weeks of training and capable of carrying out small unit and guerrilla style operations. Estimates of their overall effectiveness gives them credit for roughly 30% casualties against VC/NVA, also called the People's Army of Vietnam (Socialist Republic of Vietnam) or PAVN forces.

The second group was the Popular Forces, a part-time local militia of the ARVN. Their role was mainly to protect their local homes and villages from attacks by the VC. They were more like local security forces put in place to guard and defend critical geographical points such as crossroads and bridges against lesser organized local VC forces. They were also a form of outpost system. They were first referred to as the Self Defense Corps. Collectively they were called the *Ruff-Puffs* by American forces.

There was a third group that flew under the radar of public knowledge and this unit operated through the CIA's Phoenix Program. This program was a counter strike to the NLF, National Liberation Front, run by the VC, which was directed by North Vietnam to terrorize villages and hamlets, chiefs, and anyone suspected of aiding South Vietnam. The intent was to create a revolution in the ranks of the local population to support the North's efforts to overthrow the South Vietnam government. Assassinations were common and chiefs, teachers, doctors, and others who possessed knowledge found themselves a target. The poor villager soon found himself or herself torn between the NLF and the Phoenix Program. Assassinations and torture were the essence of this war and they ran in the background, unnoticed by us.

Phoenix sought to provide cooperation between various Vietnamese and American agencies so that they could coordinate their war on the Vietnamese people. What was the Phoenix program? A massive campaign of torture and assassination aimed at destroying what the CIA called the VCI the Viet Cong Infrastructure.

The idea was to attack the civilian Shadow government the Communist revolutionaries had set up across the country. In many places they were the ones in real control on the local level and they used their power to fight for land reform on behalf of the peasants. America's South Vietnamese allies however were a corrupt class of elitists who supported the landlords against the peasants and were hated by the majority of the populace. For the Vietnamese, the Vietnam War was a revolution aimed at land reform and re-uniting the country.

It wasn't until years later after reading some comments from a former NVA soldier that I could understand why some of the local villagers hated the Americans. The North did an excellent job of selling the concept that we, the United States was in Vietnam to replace the French and were there to continue to colonize the country and take control of every aspect of their lives. For us, the view was the exact opposite, to save this country from the overthrow by the Communist North Vietnamese. To us it was about the freedom of choice and liberties we lived by and wanted other countries to experience. It looked like we were losing that argument.

The Phoenix program had its roots in an attempt to mimic the tactics of the Viet Cong. Vietnam was quite literally a giant laboratory to study counter insurgency tactics. They called the tactic counter terror ironically since what they were doing was terrorism. In fact throughout the Vietnam War, the CIA staged false flag terror attacks even bombing theaters and blamed them on Viet Cong Terrorists. It's surprising that this aspect of the war is little mentioned given its special relevance today. However the primary purpose of the death squad commandos was to sneak into NLF controlled territory and kidnap and kill some target. They also served as an intelligence gathering agency. You could identify this group out in the countryside by the skull patches they wore on their uniforms.

Echo Company, formed late in 1969, and commanded by CPT

Graves and later 1LT Verne Seehausen took over responsibility for the heavy weapons platoon, 4.2 inch mortar and the C.R.I.P. unit. I learned later from someone who served in that unit that they were also conducting Phoenix style operations out of Trang Bang. CRIP stood for Combined Reconnaissance and Intelligence Platoon and operated mainly through insertion and behind enemy lines to gather intelligence information while avoiding detection.

Lastly was the National Police or state police which had their own stately style of uniforms typical of what you were seen worn today. They were responsible as a normal police force chartered to control the civilian population as a civilian entity, not a military one. We used them around Hoc Mon as they did not get too far from Saigon. They were helpful to question individuals we had picked up during a cordon operation so we could ensure that the personnel identification papers were up to date and matched the individual as well as for interrogation. They also assisted us in traffic control. The National Police were primarily founded to suppress internal political dissent and organized crime. Also mentioned previously was the use of Kit Carson scouts and Chieu Hoi's as supporting elements of our operations.

16. SEPTEMBER — FSB STUART IN TRANG BANG

September 1st LTC Don Green announced a change of command. 1LT Richard Wiggins from 2nd platoon was given command of Alpha Co relieving 1LT Michael Christenson and a week later 1LT R.W. "Bud" McDaniel took over as Charlie Co C.O. from CPT William Parish who was our C.O. backfilling in for 1LT Ron Hendricks who moved to HHC. This was just for a brief period, about a week, then Parish returned to his battalion duties. 1LT David Riggs reported in and took over 3rd platoon with McDaniel's departure.

Wiggins would recall later that he was called in to see LTC Don Green, the Bn Cmdr after only being in the company for 45 days and was asked if he wanted to take over a company as a commander? Wiggins asked for a bit more time as a PL before he would feel comfortable with the move.

Activity remained light so far in September and we continued to come up empty with our village sweeps and check points. We finally got some air time and conducted air assaults to the area to our northeast on Sept 9th and made four landings on Sept 13th and one on the 14th but with no results. We continued to receive occasional sniper fire in our travels, but no real enemy engagements. It looked like there was little else we could do in our current AO.

With Charlie hiding, I guess it was a good thing that we kept him from rallying his forces and striking back at us. SFC Kaplan was reassigned to Battalion. The ambush patrols continued as a daily entrée. In conjunction with our night work, we crisscrossed the entire area following information passed on by S-2 to S-3 giving us areas to search for more weapon and food caches. Every company seemed to be finding their fair share and enjoying some success toward our efforts to stifle enemy activity in the region.

On one mission where we were conducting a cordon, Partee was acting like he was at a family BBQ. We went out to surround a village called AP Binh at night. That was a company force in size. Our platoon was blocking any exits from the village to the southeast. We were spread out along the road and edge of the village, spaced out to cover a 200 meter area. SFC Partee was smoking his pipe! Smoking a pipe, get it? Now, who doesn't know what pipe tobacco smells like. And, he was talking in a normal voice like we were at a tea party. We all had to give him the elbow and told him to cut with the chatter. Now, this guy was a veteran, but what the fu** was he doing? He came back at us with the Lord will protect us. I didn't want to argue with Jesus at that point, but give me a break, Partee. We're not supposed to tell Charlie "You who, oh you who! It's the U.S. Army over here fellows, so don't come over this way." Like I said, Partee was OK, but sometimes, he was a little too "laid back" and hospitable to comfort me.

Today September 14th we got a new battalion commander and there was a change of command ceremony scheduled. LTC Don Green was being replaced by LTC Thomas F. Dreisenstok. The book on him was that he was a West Point graduate. We don't get much information about these officers. They were just another pretty face to us. There was a lot of fanfare with these ceremonies. Everyone had to dress up, that is put on a clean set of fatigues, blouse our boots (which is to tuck our pant legs into our boots), and look especially pretty while we stood in platoon formation as the band played and the officers marched by in review. Then, the brigade commander took the regimental flag and handed it to the acting commander who then gave it to his replacement and the transfer of power was complete except for some saluting, maybe pinning a medal on someone's chest, and some words to the effect of, "What a great job you did, blah, blah, blah." We were dismissed and went back to our bunkers to pick back up with whatever we were doing before the interruption.

September 15th we were airlifted to XT782065 which was miles northeast of where we had been working to conduct RIF's and weapons searches. In the process we searched the waterways looking for weapon and food caches hidden along the banks and canals. After wading through water all day we discovered 65 RPG's. They were wrapped in wax paper and hidden among the weeds and reeds at the water line. The next day we returned to the area and found all sorts of weaponry, including 39 82mm rounds, 153 RPG's, and 13,500 rounds of AK-47 7.62mm ammo and sacks of rice. Anything we found aided in our cause of slowing down the VC. Spending all day in water, especially dirty water, had its price tag. Once we stepped out of the water and got to dry ground, everyone stripped down and did a quick check for leaches. The one effective tool for removing these ugly little creatures, was either a burning cigarette or grabbing your military issued insect repellant and giving them a quick squirt. We all managed to get some digs in with each other and get a good laugh as we discovered where the leaches had attached themselves. A man's private parts seem to be a favorite haunt. In spite of tucking our fatigue pant legs into our boots, these critters still found a way into our clothing.

September 18th and our third consecutive CA into this area produced another cache find of 4,500 rounds of 7.62mm. Alpha Co under the command of 1LT Richard Wiggins had also been discovering large weapons caches in the area. The day we found our large cache on the 15th they found 122mm rounds, 75mm rounds, RPG's, small arms ammunition, TNT, 60mm and 82mm rounds, all under a bridge along a canal and got a nice write up with photos in the Tropic Lightning Newsletter, the 25th ID paper. This find was so impressive that COL Lewis Ashley, 3rd brigade commander flew out in his chopper to see for himself. We made one last assault on the 19th which turned out to be another quiet day and returned to our NL. The next two days also proved uneventful. We had watched other units being pulled out of this area, the 4/9th and 1/27th and thought our time was short also.

Ed Wales from my squad got the word to hitch a ride down to Cu Chi with the resupply convoy to see First Sergeant Johnson. Upon arriving there, he walked over to BN HQ and told the desk clerk that he was reporting in and the clerk waved him into the First Sergeant's office.

"Sit down, son," Johnson said. "I got a memo from the Dept. of the Army, who was contacted by your mother. They were told that she doesn't hear from you, that you don't write. Write your Mom," he said. "You are excused," and with that, Wales returned to the company. Later, he said, "That's my Mother for you."

We finally got orders to leave this area we had been in since the second week of July. The BN was airlifted to Cu Chi for stand down, to get cleaned up, use the showers, then draw a fresh pair of fatigues, get some of our equipment repaired, do inventory, attend to administrative matters, and do a bit of restocking of items we can't get in the field. The battalion was moved over to the control of the 2nd Brigade as well. We were there for several days, then were sent to FSB Stuart by truck convoy, located on Hwy 1 just outside the southside of the village of Trang Bang and up the road from the division basecamp at Cu Chi.

Alpha Co's C.O. Richard Wiggins recalls, "If my memory serves me right, the entire 2/12th was moved to FSB Stuart and we relieved the 1/505th BN, 101st Airborne unit. They were leaving on Hueys as we were arriving and I had about a two minute briefing with the company commander whose positions we were taking over. CPT James Duggin, D Co, was replaced by CPT James Ellis."

My new platoon leader, David Riggs was really keeping me busy since I got the promotion to buck sergeant. I don't know if he was testing me or just wanted stuff done. I was happy to oblige him. At FSB Stuart there was a battery of 105mm's and later, D Battery 3/13th, a pair of 8 inch self-propelled howitzers located here in October. The morning of September 23rd, Companies A, B and D moved up TL-6A, a small dirt secondary road about six clicks due north and established a night laager which was used by the battalion.

While the rest of the battalion was to our north, Charlie Co was assigned to FSB Stuart for laager security and to protect a bridge just down the road about 200 meters from the FSB. McDaniel's plt was assigned to the Hwy 1 Bridge while the rest of the company settled into the FSB perimeter bunkers. It was important that this bridge was protected from sappers. It sat on the MSR between Cu Chi and our bases at Dau Tieng and Tay Ninh.

Since we had arrived at Stuart and with the rest of the Battalion to the north of us, our job was to keep the immediate area secure. So our patrols for the first two weeks were limited to the area immediately around Trang Bang and the area to our north and east. During the night we secured the fire support base near the Buddhist Temple which was situated next to the main road and on the southern edge of Trang Bang village. During the night when we had somebody on duty guarding the bridge, their job was to throw grenades into the water to discourage sappers from swimming down the river and planting explosives on the bridge. To aid us in the detection of movement, we had a cage of three white geese located on a sand bar upstream of the bridge about 50 feet. They started honking if anyone approached them.

Located here, we were in the main corridor for the VC to funnel troops and supplies toward Saigon. To our northwest was the Boi Loi Woods and to the northeast, was the Hobo Woods, and across the Saigon River, the Iron Triangle. To our east lay the Filhol Rubber Plantation, located north of Cu Cui. These were all heavy areas of enemy activity. To our northwest was the village of Go Dau Ha sitting beside Hwy 1 and where FSB Hampton was located. It was just outside of our new AO and the responsibility of 1st Brigade.

Besides sharing the FSB with the artillery guys, we had A Troop, 3rd Squadron, 4th Calvary operating out of there on occasion as they patrolled Hwy QL-1. They were the tank troop, using M48 tanks, nicknames *Big Boys* and APC's. As the ground began to dry out, this unit began to venture out on operations off the main road. Some areas were drier than others. The area south of Hwy 1 remained a quagmire but to the east and north, the footing was more willing to carry the weight of the tanks, almost.

It took a few days for the mail to catch up with us since our departure from the Hoc Mon area to our new location. Resupply, the guys tasked with running convoy's between the division base camp and FSB Stuart, showed up almost daily with trucks loaded with provisions, food, ammo for the artillery and troops plus the mail. It's not long before someone shouted out "mail call" and we quickly assembled around the man toting the red mailbag. It's Sydney Fowler and Hiram Marziano, who tagged along on this trip. Sydney grappled with the mailbag, then pulled out a handful of envelopes and began to shout out names as he read the addresses. I was in luck and had letters from Mom and Dad, Pam and one from Berry Blackmon, my neighbor from high school. I moved away from the group and opened the letters carefully reading their contents. Berry wrote, "Hi Arnold, I wanted to let you know that Vicky Vaughn and I are engaged and will be married on October 26th. We wish you could be here, and we hope that all is well with you and that you are safe and sound...etc." It was nice to hear from him. I found it ironic that the girl he's marrying was someone that I had briefly dated when I was 15 or 16. I wrote him a brief note then got back to my duties with my squad.

The battalion continued to concentrate all our search efforts into the area to our northeast. Daily we had one, two, and sometimes all three companies out doing RIF's, jumping from one village to another. During the remainder of the year we would visit Xa Sa Nho (1), X. Rung Cay, Ap An Binh, Xa Tu Duan, Bao Me, Gai Lam, and then eagle flight across the Trang Bang River to Cau Chua. Next it was Xa Bau Tram, then another eagle flight to the Citadel at Xa Rang (1). Back and forth, we would crisscross trying to catch the enemy napping.

The first few weeks of October we found out the going here will be much different than down around the southern reaches of Hoc Mon. The enemy was very active and we were in daily contact with him. October 1st through 5th we found ourselves in small but brief clashes with the VC.

The night of the 6th we were sent south on an AP toward X. Rung Cay. We arrived at our checkpoint. The platoon set about its task of placing Claymore mines and in this case we were told to also put out trip flares. SGT's Conlin and Keatts and their squads were working with their guys on setting up the mines. I had my guys placing a few flares on the approaches to the AP site. The flares were painted a flat green color. I had picked a location for one. I unwound the trip wire which is a very thin wire like a piano string. I tied off one end to a branch and stretched it across the trail. As I lifted the spring loaded firing pin to slip in the safety pin, it broke off in my hand or somehow it slipped out of my grip. Either way, the result was not good and I set off the flare while holding it. It burned three of my fingers from the intense heat it generates. It burned magnesium which puts out a very bright white light. I couldn't do anything but take cover and wait for it to burn out. If Charlie didn't know where we were, he certainly did now.

We were not going anywhere so the order was given to dig in. We broke out our trenching tools and started to dig. We made firing positions deep enough for several of us to fit into each one. About 0100 hrs we got the familiar BLOOP, BLOOP, BLOOP sound of a mortar round leaving the tube as Charlie began to target our position. He was a skilled marksman and managed to walk his exploding HE rounds right across our positions. Fortunately, no one was hit, but I found myself trying to get as low as I could in that foxhole as each exploding round came closer and closer to our position. I was getting pretty nervous when he stopped his mortar attack. I could visualize a round landing in our foxhole. The rest of the night was quiet and in the morning I had Doc Green tend to the burns on my hand. Not a good night for me.

Word came down from Battalion that everyone was mandated to wear flak jackets on operations. Why now and why us, was not explained. They can't stop bullets, but they do offer some protection. Most of the mech guys riding the APC's or tanks wear them, but they are just too awkward to wear when we're humping all day long. After trying them on and off for several weeks, I decided to dump mine. They were bulky and hot and I didn't need the security blanket. No one was challenged by the officers for not wearing them by then.

Each patrol was like the other. It was like a walk in a large park, only you're carrying weapons. You walk, stop, start again, and stop again, like a large centipede. You were usually in two columns, each soldier keeping some distance from the other to minimize damage in case of an explosion or burst of AW. Paralleling each column were your flankers, usually two infantrymen walking about 25 to 50 meters out and away from the main column. This was your eyes and ears to each side of the main force. They were like advanced scouts looking for danger and anything suspicious.

If it was a company operation, the front of the column was a rifle platoon comprised of 25-30 men carrying M16's and M79 Grenade Launchers in three squads and the fourth squad was a heavy weapons squad comprised of two M60 machine gunners and their support crews. There would be an assistant gunner, and two ammo bearers per gunner. Each squad was led by a sergeant and each platoon had a platoon leader, usually a 2LT or 1LT lieutenant with his RTO and a platoon Sergeant, also with an RTO. Behind the lead platoon was a second rifle platoon followed by the HHC group. The HHC was your company commander, his RTO, an F.O. for artillery and sometimes a F.O. for tactical air support. And finally, there was a battalion RTO to keep us connected with the higher command. Following this group was another rifle platoon. If the company was at full strength, which would be unusual in Vietnam, the company strength would be around 80-100 men.

A quick overview of the weapons a typical infantry company used started with the M16 rifle which fired a 5.56mm 55 grain cartridge and could fire either single shot or by throwing the selector switch could be turned into an automatic rifle with a rate of fire of approximately 800 rounds per minute. The M60E4 machine gun fired a 7.62mm 150 grain NATO cartridge with a cyclic rate of fire at 500-600 rounds per minute. Any more than that, and the barrel would eventually melt from the heat. Machine gunners carried a Colt .45 model 1911 pistol for personal protection. The last weapon used was the M79 Grenade Launcher, a breach loading single shot weapon which fired a 40mm grenade that exploded on impact with a killing radius of 5 meters with a range up to 400 meters. It was nicknamed the "blooper" because of the sound it produced when fired. A few guys carried personal weapons, which was not supposed to be allowed but somehow they had them anyway. John Mikita carried a pump action shotgun when he was on point.

We were in no hurry to make mistakes or take any chances. We moved cautiously but with purpose. Every so often we took a break and fell out for 10-15 minutes. Later on, it was time to grab some lunch. 1LT McDaniel ordered a stop where we were in the search and we broke into whatever C-Ration we grabbed prior to leaving the FSB. So, I opened up my cardboard box labeled C-Ration and printed under that was Spaghetti. I pulled out the can of Spaghetti, and under it, was a tin of crackers and a tin of Cheddar Cheese. Next to these items was my dessert, a can of fruit cocktail and a napkin, along with a spoon. I had all that I need for lunch. My stomach couldn't handle anything too spicy anymore. I guess it was the nerves and constant tension we were under. I was popping antacid tablets to try and calm down the burning sensation I was feeling. I got by on eating the canned fruit from the C-rations most of the time, along with the crackers and bread. Some of these meals had a lot of fat or grease in them and were unfit to try and eat cold. I passed on the spaghetti. Although we had stopped for lunch we were constantly aware of any activity around us. We kept our sentry's out on the flanks and the point and rear guard remained vigilant.

After a while, we had gotten quite creative in changing up our menu's when it came to our box lunches or dinners. Obviously, I'm still describing the variety of meals prepared using C-Rations. First of all, we had a handy little tool called a P-38. This was a small can opener that came with our C-rations. It was collapsible, about an inch long, and fit in our pocket. That was a very useful item to have on hand as there was no getting into any sealed can of food without one. Some guys, like me, carried a bayonet, but

I wouldn't say it would work well to open a can unless you were totally desperate.

We had learned to combine canned white bread with a packet of cheese and some spiced beef which had a bit of sauce to it, and by layering the ingredients back into the original can which you carefully remembered to not completely remove the lid with your P-38. You then took the can and closed the lid over these ingredients and put it back into the C-Ration box and closed the lid and set it on fire. After the box completely burned, you had a nice and hot pizza waiting to be eaten. Most food could be heated in the can using a small ball of C4 explosive. Everyone carried a small chunk of the stuff in their pocket. It burns very hot when you light it. It's like using *Sterno* for cooking. You just break off a piece of C4, roll it into a ball the size of a very small marble and touch a match to it. You prop the can you want heated over the burning C4 until the food is heated.

Every so often we crossed paths with the 75th Rangers (LRRPS) in the field on patrol. They operated out in the open and alone. Their mission was to see and not be seen. It was a very difficult job, much like a Navy Seal or Green Beret. They were a small team and usually less than a dozen men. They were advanced scouts on foot and their primary mission was to collect intelligence data on troop movements and locations. During one of these moments in time when our paths crossed, we managed to beg them to give up some LRRP (slang is *lurp*) rations they had. These are great and really break up the routine diet of C-rations. LRRP rations are actually freeze dried foods, even ice cream that you add water to, or not, and *voila*, it's like real food. Our supply of LRRP rations did not last long, but were really enjoyable while we had them. They had choices like beef stroganoff, chili con carne, fettuccini, freeze dried ice cream, and we had to eat ham and lima beans and spaghetti.

Bravo Co under the command of CPT Allen Wissinger air mobiles east of Pershing on October 7th and walked into a bee's nest at Tam Dinh. They suffered six KIA's in a vicious firefight, including medic Ramos Flores and eight WIA's. They joined up with 1LT Richard Wiggin's Alpha Co earlier in the day who was also operating in the area. SP4 Larry Fontana was acting as a sergeant that day — a 'acting-jack' sergeant — and was on point with Bravo company's 2nd platoon led by 1LT Howard McKinney, when they entered the village of Tam Dinh: "The biggest fight we were in was four klicks east of Pershing October 7, 1968. Bravo and Alpha went out to make contact with a large force of NVA spotted there the past evening," recalled Fontana. "Good as gold Alpha abreast to our left. My platoon made initial contact in the middle of a village and did some damage — but we got a bloody nose," said Fontana.

Bravo Co beat back the NVA first, and then, "We could have got out without a scratch … There was a window of opportunity after bloodying their noses, but an officer made us stay. Fought off and on all day and dug in that night with two companies from the 101st Airborne Division (one KIA and 13 WIA's)," said Fontana.

In addition to the Airborne, there were the two companies of the 2/12th — Alpha and Bravo. "NVA hit us with the kitchen sinks and came charging in. We were fighting from open foxholes and one strand of concertina." Fontana continues, "No food or water, but plenty of CLAYMORE MINES!!! Saved our bacon without a doubt. We had Claymores all over the place and broke up the North Vietnamese Army's 101st Regiment after three charges against us," stated Fontana. "Probably lasted an hour, but seemed like eternity."

After a day long battle in and around Tam Dinh, CPT Allen Wissinger's B Co and CPT James Ellis, leading D Co, which was flown out to support B Co, with two additional companies of the 3rd Bn 187th Infantry from the 101st Airborne, coiled into a defensive position northwest of the village. Enemy losses were 55 KIA, 46 possible KIA, 18 KIA by artillery with another possible 11 KIA.

The following morning at 0805 hrs, the companies received 35 rounds of 82mm mortar fire, which was followed up by a ground attack. The result was 72 NVA KIA, 26 possible and one POW.

Two days later, on October 10th, Alpha Co got into it over at Lam Vo with, now newly promoted CPT Richard Wiggins the C.O. They were ambushed and suffered four KIA's, three received the DSC later, Stanley Denisowski, Baynes McSwain Jr. and Michael Randall, Sr. SP4 Eugene Handrahan was wounded and could not be reached under fire. He disappeared from the area and was reported as MIA. His body was never recovered and is still listed as MIA. During a series of running battles that started on the 7th, Alpha company was approaching a hedgerow when an enemy machine gun opened up on them. Grenadier Handrahan, who was walking point left-flank, and two others were hit. The two other G.I.s were under aerial observation and determined to be dead. Gene yelled that he was hit, and repeated attempts were made to get to him. After a soldier was killed trying to get to him, the squad was ordered to pullback in order that the hedgerow could be bombed. Wiggins made that call late in the day. It wasn't until the next day that the company went back in. They found the bodies of the three G.I.s, but no trace of Gene Handrahan was found. A large bomb crater was at Gene's last known location. He is the ONLY MIA that the 2/12th Infantry Regiment had in Vietnam.

On the 11th Charlie Co made contact as one of our flankers was hit with small arms fire. 1LT McDaniel called for a LFT and TAC air. Crawling nearly 100 meters only inches below crisscrossing fire from four enemy machine guns, Doc Dennis Sheppard worked to rescue his critically wounded buddy. It was difficult to pinpoint the location of the four rattling machine guns. As the company's senior medic, Specialist 5 Dennis R. Sheppard of North Hollywood, Calif., began his crawl from a point near the center of the troop column.

Doc was joined by rifleman PFC Dennis R. Buckley of Burbank, California. Together they maneuvered under withering fire.

"We moved in a zig-zag pattern for about 50 meters," said Buckley.

Reaching the injured man in knee-deep grass, Sheppard decided more help would be needed. As Buckley returned for assistance, the unarmed medic began administering first aid. Gunships roared in overhead to cover the evacuation. Firing within 20 meters of the wounded soldier, their rockets rained tiny bits of shrapnel on the struggling medical corpsman.

Moments later, three more *White Warriors* reached the scene. Carefully moving the still-conscious man across 30 meters of exposed area, they reached the cover of a hedgerow. A Dustoff evacuation chopper had been summoned.

"But the situation was critical," said Sheppard. "Thus, it was necessary to have the injured man extracted by gunship."

Within minutes the medic and his patient were inbound to the 12th Medical Evacuation Hospital in Cu Chi.

"A lot of credit was due to the crew of that gunship," said Sheppard. "The door gunners worked like experienced corpsmen. One of them managed the intravenous bottle throughout the flight."

The other three riflemen who aided in the rescue were SGT Arthur Hood of Churchrock, New Mexico, SP4 John Houck from Farokoa Queens, New York and Lynwood C Keatts of Gretna, Virginia. Both Sheppard and Buckley were awarded the Bronze Star for this action.

Once the troops had returned back to the main formation, everyone in the operation just stood their ground and waited for the FAC to show up on station to direct the incoming aircraft on where to place their bomb loads. If the action was heavy with the enemy, this process can be done quickly by the Air Force. The FAC showed up in his North American Rockwell OV-10 Bronco. It was a unique aircraft that had a split boom tail, much like the P-38 from WWII. Soon after, we were hearing a pair of fast movers, Phantom F4's, streak by our location at around 600mph. They banked sharply to the left then slowly peeled off to get a layout of the land as they listened on the radio to the FAC. He told the pilots that the target was a large grove of trees marked by a small oxcart trail.

The fighters made their target run spaced about a ½ mile apart so they didn't fly into the shrapnel from the 250lb bombs they were dropping. After several passes, we watched and felt the earth shake from the impact, then saw the dark black mushroom cloud from the detonation slowly rise above the tree line. We waited about 10 minutes after they had dropped their ordnance, then slowly tested the landscape by probing forward to see if we could draw fire. Nothing happened so we slowly but carefully swept the area but didn't find any bodies. How Charlies get away at times was a mystery.

The 25th ID had other units from the 1st and 2nd Brigades working this area as well because of the size of the enemy forces we were in contact with. The 2/14th Golden Dragons, 2/27th Wolfhounds, 4/9th Manchu's and 1/5th Bobcats all pushing the 88th NVA Regiment and the 272nd VC Regiment back around as well as our own 2/22nd Triple Deuce and 3/22nd Regulars. It was rare for us to work directly with these other battalions that were not part of our brigade. They also had their own TAOR (tactical area of responsibility) to deal with, but at times we combined operations, especially when someone made contact with a large enemy force, When that occurred, we quickly moved to assist either directly with reinforcements or as a blocking force attempting to bottle up the enemy so he couldn't escape from the trap.

I knew we had worked with many mechanized units, 3/17th,

2/34th Armored, 1/5th, the 3/4th Calvary and our own brigade, 2/22nd. I would have to say that I worked more with these units during the last half of my tour than the first half and the most likely reason was the area we were in and that it was the dry season now. This was especially true when we were operating in the Hobo Woods, Boi Loi Woods, and northeast of Trang Bang.

What was our tactical plan of attack against the enemy? This was the first war that the U.S. was fighting in which we had no clear enemy that could be identified mainly by uniform. Our adversary could be the local farmer, vendor in the market place, a Mama-san, a Papa-san, or even a child. He could be a regular Viet Cong soldier living in the country or an NVA soldier. He could fight alone or in groups. He could also band together to fight as a large force from platoon strength to regimental strength.

Why we had chosen a strategy which did not allow us to retain control of the ground we walked and flew over is unclear to me. The old adage that says "keep your friends close and your enemies closer" might have worked here because they would have been under constant watch. Later, I learned that at the beginning of the war it was decided that we would fight a war of attrition, hoping to defeat the enemy through large loss of life. So, every battle was gauged by body count. Therefore, taking control of land was not the objective decided by the higher ups. We failed to learn that the Asians are very patient and time is more important than life when it comes to achieving their goal.

When I arrived in country, we had been using a tactic of using large bases to operate out of during the daytime and staying in these at night. For the most part, this was true, unless we were conducting RIF's or combat assaults based on intelligence, then we could be operating out in the field for weeks. We did go into defensive positions at night, much like the covered wagons of the 1800's. It is very difficult to overtake a night defensive position when you have your firepower concentrated into a tight circle and you are generally protected by a FSB containing long range artillery. If FSB's had 105mm guns they could cover an area around seven miles from their location.

One of the major innovations to come out of the Vietnam War, was the creation of the fire support base. This was due entirely out of the fact that we had no defined battle lines other than the DMZ at the top of I Corp which separate the two waring fractions.

The fact that every unit was mobile and classified as a maneuvering unit meant we needed some means to provide fire support and this could not be accomplished through major basecamps. There were often times when that support needed to be in proximity to the battlefield. By late 1966, the usual procedure was to establish fire support bases containing all the elements found in a basic battalion within the FSB if it were to be in a location for an extended period of time and the makeup of those elements were defined by the size of the artillery batteries and supporting troop elements. For example, headquarters elements, medical facilities, an area for a PZ.

The fact that the early use of these FSB's were battle tested early by the NVA and VC forces and proved to be highly defensible was a positive. As time would prove, they were also an inviting target of the enemy and they paid dearly whenever they attempted to take and overrun one. The defensive tactics used in the creation of the FSB with its bunkers, foo gas defensives, the later invention of the Killer Junior air burst 105mm round, use of flechete rounds, quad .50 cal gun carriages all led up to destructive firepower the enemy could not overcome. They also served a dual purpose of giving ground troops a place to operate out of and at the same time, provide defensive support for the gun batteries. At times, they even served as a decoy to lure the enemy into attacking them, a mistake made too often, and the losses too great.

The country of Vietnam is not big and it was easy to have a

network of FSB's equipped with 105's, 155's, 175's, and 8 inch guns ready to respond to a call for help. The 105mm fired a 42 lb round about seven miles. The 155's could fire a projectile weighing 95 lb nine miles and 175's fired a 174 lb round which could go out around 20 miles. The M110 self-propelled 8 inch howitzer could cover close to 14 miles with its 200 lb projectile. The 8 inch gun is considered the most accurate field artillery piece in the Army. The majority of the artillery batteries in Nam were equipped with the 105mm howitzer. They were easy to airlift by chopper wherever they were deployed to.

With FSB's dotting the landscape, it made it easier to maneuver wherever we wanted to and we could also engage large forces directly because of our ability to call in for close fire support either by artillery, gunship, or tactical air support from Tan Son Nhut AFB. The fighter cover was made up of F-4E Phantoms, F-100 Super Sabres, F105 Thunderchiefs and others. The A1-E Skyraiders (prop driven) were flown by the South Vietnamese Air Force although the U.S. Navy had these as well.

Our gunships were the UH-1 Iroquois or AH-1 Cobras. The Cobra was superior to the UH-1 for accuracy and bringing on the heat. The gunships had twin miniguns (Gatling guns) which fired 3,000 rounds per minute plus twin 16 pod 2.75 inch rocket launchers. All weapons were controlled by line of sight guidance. We also had Spooky, which was an AC-47 that had 3 miniguns mounted to one side and used for ground support. Later on, this same gun mounting was used in an AC-130. It too had a nickname, which was Puff the Magic Dragon. For the jets, napalm and HE bombs were the first choice for us on the ground and when needed, they could drop WP.

So there you have it. The ground forces would continually sweep the countryside looking for any enemy troops caught congregating together and eliminate them before they could organize any large assaults against our forces, NL's, or basecamps. Charlie and the NVA, no matter how hard they tried, would fail time and again to gain the upper hand against the allied forces. Every encounter we had with them, they got their asses handed to them. Yet, they were bound and determined to demoralize and defeat us. In the end, in spite of newspaper articles and books written about this war, the U.S. was not defeated. We ended up signing a cease fire with North Vietnam in January 1973 and departed from that country on our own terms.

By April 1972, U.S. troop levels had dropped to 69,000. As we were withdrawing troops, the South Vietnamese Army was heavily engaged with the NVA in the Eastertide Offensive where they held their own. Both public and U.S. Congressional support was rapidly diminishing at this point. By August, all U.S. combat troops had left the country. The fact that the U.S. Congress failed to fulfill our agreement to keep SVN supplied with money and equipment that SVN lost quickly to NVA. It wasn't because of their lack of will, it was because the U.S. Congress deserted them and didn't do what they had agreed to do after the signing of the Paris Peace Accords which were signed on January 27, 1973. On June 19, 1973, Congress signed the Case-Church Amendment which forbid any further U.S. military involvement in Southeast Asia. When the NVA realized we officially ended all support, they quickly violated the Paris Peace Accords and accelerated their incursion into South Vietnam knowing they could not be stopped. The bottom line was the Russians and Chinese kept the NVA supplied, and we let our SVN allies down.

The war was left in the hands of the South Vietnamese Army and it was quickly determined that they could not stand up to onslaught of the NVA fighting alone and without supplies and financial aid. By 1975 the outcome of the war was inevitable and on April 30, 1975, they surrendered to Communist North Vietnam.

Vietnam had all the comforts of home to me. For sure, outdoor camping and the opportunity to sleep on the ground was a highlight.

Preparing all your own home cooked meals except for the times when we got hot breakfasts or dinners. I should point out that the Army did a damn good job of trying to make all of this bearable. Good food kept the spirits up and we truly didn't live on C-rations alone. Breakfasts were not always available and when they were, you stayed away from the eggs, powdered or otherwise. We didn't get to raid the refrigerator at night either. Anything we felt like consuming had to come from our goody packs, or from C-rations if we missed chow time.

When it came to keeping ourselves clean, that was another story. We sweated a lot, both from the heat and from the fear of battle. You can't believe how much water you will consume during a firefight. The jungle fatigues we wore sometimes lasted 10 days or longer before we got a chance to change them for a fresh set. Everyone tried to take an occasional sponge bath or used some type of jury rigged shower stall to wash the dirt away. It felt awesome to get that refreshing shower. Sometimes you just took one when you could.

We had a few operations with multi-unit configurations. For the first time, we worked closely with both the *Aussies and ROK's*, another name for the Republic of South Korean soldiers who look as tough as nails. Our joint operation took around a week to complete before we broke up and went our separate ways. There even was a time where we crossed paths with the 101st Airborne. They appeared to really be loaded down with equipment with their backpacks, something we rarely did or used.

On October 18th at 1100 hrs C Co received SA and RPG's from an estimated 8-10 VC. Our platoon returned fire with all organic weapons (what we normally carried in the field) and LT McDaniel called in artillery fire. "Focus 27 this is Flame Charlie Six requesting fire mission, enemy squad, estimated to be 8-10, location grid XT524599, one marker round—Willie Pete. Will adjust fire, over."

"Flame Charlie Six—Focus 27 Roger. Fire mission Grid XT524599, one marker round, Willie Pete, over."

The marker round whistled overhead and hit 100 meters to the left of the area where we spotted the enemy. The artillery fire needed to be corrected.

"Focus 27, Flame Charlie Six, range correct, adjust fire right 100. One round HE, over."

"Flame Charlie Six—Focus 27 right 100, one round HE, over."

In the distance we could hear the howitzer's *report* as the gun fired and within a few seconds, we heard the round whistle overhead and strike the target area with a *KAARUMP.* There was a bright flash with the explosion, then a plume of dark gray cloud of smoke enveloped where the round landed and the whistle of shrapnel. There was no echo or lasting effect with the detonation, just a loud noise that was heard.

"Focus 27, this is Flame Charlie Six—six rounds HE, fire for effect, over."

"Roger Flame Charlie Six, firing for effect, over."

Once again, in the background we heard the battery of 105mm reporting, then heard the incoming rounds fly overhead and envelope the wood line where we last had visual contact with the enemy, *KAARUMP, KAARUMP, KAARUMP, KAARUMP…KAARUMP, KAARUMP.*

After a pause, McDaniel repeats:

"Focus 27, this is Flame Charlie 6, Fire for effect, over."

"Roger Flame Charlie 6, Fire for effect, over."

There was a second, then third volley of HE rounds, *KARRUMP, KARRUMP, KARRUMP* striking the area. The dark gray cloud of spent powder and dust increased in size. The gray clouds of expended powder curled upward into the sky, at odds with the white cumulus clouds and blue azure sky.

"Focus 27, Flame Charlie Six, cease fire, end of fire mission, Well done, over."

"Roger Flame Charlie Six, end of mission, Focus 27 out."

We waited several minutes then cautiously swept the area. The enemy had disappeared into the vegetation. The results were negative U.S. casualties. VC losses were unknown.

The order was given to head back to base, so we slowly worked our way back, carefully moving, scanning and pausing to ensure nothing was left to chance and finding ourselves flat footed and pitted against another enemy patrol. We stopped to take a ten minute break and regroup. I scanned the ground looking for a place to sit and relax. I didn't see any creatures lurking as I stretched out, but that quickly ended. One thing there was an abundance of, were crawling things. Soon I saw black ants roaming the earth in search of what, I don't know. They spotted me and soon I had ants exploring my fatigues, but by now I don't care. It had become part of the lifestyle here and they don't bite. Soon they were happy and departed for the cover of the holes in which they came out of. This routine happened all the time during the dry season.

In covering things that were on the ground everywhere in Nam, this included snakes as well. Everyone had a snake story or experience while in-country. I do not recall getting any instructions about snakes, other than the two step snake story that was passed around. To us, it was a green bamboo viper but there were other species that may also fit this description. American soldiers during the war in Vietnam called it the *two-step viper*, in the belief that its venom is so lethal that if it bites you, you will fall dead after taking just two steps. That's an exaggeration, but the bite of the many-banded Krait is astonishingly potent. The venom is a neurotoxin, which means that it disables the victim's nervous system—like yanking an electrical plug out of the socket. Death comes when neurotransmission ceases. With no instructions to breathe, the muscles of the diaphragm are stilled, and the victim asphyxiates. There is no way of knowing how many grunts died from snake bites due to how medical records were filled out.

My own personal experience was during a RIF mission out in the Michelin Rubber. Our company had stopped to take 10, and we set out our sentries. Most of the guys were just hitting the deck where they stood to get a drink or light up a smoke. For me, I needed to relieve myself, as in #2 so I ventured out into the brush a bit and dropped my fatigues to get to business. About the time that I was going to take care of business, a snake slithered right between my legs. It was black and about three feet in length. I do not know what it was, but that ended any chance I had of finishing what I started out to do.

The company arrived safely back at the base and that evening after chow, there were a number of clouds in the sky. At times this country gave us beautiful sunsets to gaze at. Tonight was no different. The clouds ranged from dark gray to various shades of gray and whitish silver with a few actually taking on a purple tone look to them. As the sun got lower on the horizon, the clouds began to transform into shades of pinks and oranges highlighted by blues and purples with bands of yellows. It would have been worthy of a Claude Monet painting.

I decided to check on some of the boys over in 1st and 2nd squads. McInvale was listening to AFRTS on the radio. He was a huge fan of *Chickenman* and couldn't wait to hear this week's episode. This was a classic which started in 1966. Each episode started with *Now, another exciting episode in the life of the most fantastic crime fighter the world has ever known... Bak, Bak, Ba BAAKKK, CHICK-EN... MAAAAAN... He's everywhere, He's everywhere.* It was a good show and even I chuckled at the antics of this character at times. Vernon Becker and Darrell Kuhnau, both blond haired, were also lending their ear to the show. I decided to listen for a bit before I moved on.

I spotted SGT Conlin who had 2nd squad and stopped to chat for a few minutes, asking him if he had heard from home and how was his squad doing. I really like Conlin. He's quiet and unassuming and has a great smile and one of the few guys who wears prescription glasses. He knows how to relax and relieve the tension of the day. We talked for a few minutes about today's operation and how the new guys performed and if they made any mistakes and what we can do to correct them. I headed off to see SGT Keatts in 1st squad, stopping to say hello to Quintanilla, a Hispanic kid from Texas, who had a real edge to him, but I liked his toughness. He did not talk much, but knows how to pull a trigger and is one of the grenadiers. Keatts had been with the platoon for, well, going on four months now. He was a small framed kid like the rest of us, and hails from Gretna, West Virginia. He had blond hair and narrow facial features. He was good at his job. We shared a few jokes as I pulled out a cigarette for a smoke. The topic of talk turned to the Zippo lighter I was holding, one of over a million produced. I'm not sure they all ended up in Asia, but they were all over the place. Their motto is *Works first time, every time*. Each one carried by a soldier has a different engraved message or emblem on each side of the lighter. They are a collector's item in present times. I still have mine. It was time to head back to my squad and check to see who would be taking first watch on the line before I turn in and try and catch a few winks before my watch began.

On October 20th at 1406 hrs the company was in the vicinity of Round Lake, a landmark to our east, after having left FSB Stuart on a RIF. We left the wire around 0830 hrs. We were in standard company formation when we were ambushed by an estimated squad of VC. The VC engaged us with a Claymore mine then withdrew. One of the things we had been schooled on repeatedly was to pick up all weapons etc. after a firefight. This was not always possible to do. The result of not doing this was finding these weapons being used on us at some point in the future. The enemy strike resulted in four U.S. WIA'd. A Dustoff was required. SP4 Arturo Quintanilla, 1st squad and one of our grenadiers, was nicked up from this and was flown to Cu Chi to 12th Evac. The next day he returned and was walking around with a patch over his eye. He would be camp bound until he healed.

Back at the bridge on Highway 1 near Trang Bang, the Battalion gave us a new toy to play with. It was an electronic box on a tripod and was actually a ground radar device. It could detect movement up to 1,000 meters away. The platoon and yours truly got a training session on it and left it in our hands to use. Over the time we were there, I called in several fire missions based on what the radar was indicating. The unit was good at detecting targets, usually multiple targets and distance to their location. The guideline we used was simple. Because of a standing curfew, no one was supposed to be moving around at night and anyone who was had to be considered the enemy force.

After one particular fire mission which I called in for movement around 2200 hrs, I asked for six rounds of 81mm HE on azimuth 195, about 800 meters out. Around 2330 hrs, I get orders from LT Riggs to go out and do a recon of the area the next morning. The LT told me, "You called for the fire mission, so take your squad and go find out what you hit."

I said, "Yes sir, just my squad?"

"Yep," he replied. "If you get in a situation, use your head."

The next morning, I alerted the squad that the six of us were going on a recon and to saddle up. We grabbed a PRC-25 and radioed the C.O. we were moving out. We had about 800 meters of rice paddies to navigate through. The fields were flooded and in some stretches, the water was up to our chests. This was very deep for rice fields. I had taken an azimuth of the direction we had used for the fire mission last night. The compass reading was leading us to an area of scattered huts and a tree line. Maybe I shouldn't have called the fire mission now that I saw this was a modestly populated area and likely there were civilians wandering around late.

As we turned and changed direction, we were about two hundred meters from the edge of the paddies. Wales and I continued to close in on the huts. Zimmerman, Toto, Sovey and Tostado trailed behind. When Wales and I got to the hundred meter mark, we started receiving sporadic fire from the trees. Well, that dispels the notion we had been engaging innocents last night. We had no cover, and believing in the concept of discretion being the better part of valor, we pulled back to a safe distance and reported back to company with our status. Since we were a small recon unit of six men, we were ordered to pull out and that the company would deal with the situation using a larger force. Taking on the enemy with six men was asking for trouble, maybe at night ok, but not in daylight unless backed into a corner. Just one guy getting hit left only one rifle for covering fire.

Charlie Company seemed to be the lucky one. The rest of the battalion, A, B and D companies had some really heavy action while we were stuck pulling security duty at the bridge. The casualty count was high but not severe compared to the licking that the NVA took. It was only a matter of time before it would be our turn to be engaged with a large enemy unit.

Every day that we are out sweeping the hedgerows or villages, we experienced sniper fire or were ambushed. The unit was used to it and we hit back hard. There was one village in particular that was very pro VC and in spite of our frequent visits to this location, we found no suspects. We were ordered by battalion to return to this village. Upon arriving and engaging in the usual activities, i.e., conducting a thorough search of the village, we got word to burn the village. I guess the higher ups had had it with the lack of cooperation and unwillingness to help the allies, so we were told to make this village an example and burn some hooches. The grunts were not happy about doing this. The farmers were poor and this was all they had so it was a sad day for us in some ways. Others thought about comrades we had lost, possibly as a result of these villagers. We completed our task and left knowing this was not the end of it. Some might say this was an act of evil or a war crime, but no one was killed by this incident and we made our point. The area around this village had a very high incident rate for booby traps, enemy sniper fire, and RPG attacks.

That evening, Hill, Corum, Price, Buckley, Wales, Christianson, Keatts, Toto and myself gathered around after we got our gear squared away and after dinner. It was time to take some pictures which we did once in a while, especially when a PX run was done to get film. Each time it was a different group in the photo. No one packed a camera around on patrol, although it would have been great to record some of the action we saw. After we ran out of camera film we sat down around the bunker to listen to the radio from the *World*. We tune into AFRTS coming from Los Angeles. The DJ was playing a Beatles tune: *Hey Jude, don't make it bad. Take a sad song and make it better. Remember to let her into your heart, then you can start to make it better*. After some time the radio stops working. It was time to change the battery. We got good at applying the used batteries from the PRC-25 radio to run our radios that we got from home or from the PX. We didn't have access to regular batteries so this was a good solution.

Conversation ran the list of topics. Everyone wanted to talk about girlfriends since nobody was married except Toto, I believe, at least in this group. We talked about what we were going to do when we got home, and what our favorite pastimes were. We were cautious as we lit up a butt or two, making sure Charlie could see neither the flame of the match or the glow of the cigarette to be used as a target. We held the cigarette by cupping it so the lit end was pointed toward the palm of our hand. The routine was the same every night, except when pulling ambush patrol. The players, that is, the guys involved in the banter, may change often, but the ritual was always the same.

Conversation died down around 2200, but before we switched off the radio, we listened to one more tune. It was Otis Redding singing *Sittin on the Dock of the Bay, watch the tide rolling away*... It was time when we needed to start our *night watches* and we drew straws to see who would pull first watch and who would have the 2nd, 3rd and 4th shifts. I took the first watch and started scanning the tree line using the Starlight scope. There was nothing out there but I did not give up. In the distance, I could see a flare ship working the night sky. Someone was in trouble and I could see the faint trails of tracer rounds ricocheting off into the sky as they struck the ground. I could make out the distant sounds of explosions, probable artillery fire. The action was coming from our east. The next morning we heard that the 1/5th Bobcats got into a scrap with Charlie in their NL site.

October 21th arrived with clear blue skies. Charlie Co was to sweep southeast toward X Rung Cay (2) village crisscrossing the Trang Bang River. The company rallied into formation and we moved out from FSB Stuart. Third platoon had the point, but it was second squad's turn to take the lead. The village was about two clicks away, about 2,000 meters. This was a RIF patrol. The company moved along quietly and we reached the river and searched up and down for a crossing. After a few minutes and several aborted attempts to cross, we rigged up a line to the point man for safety and sent him across. The water surface was lazy and the current was not too strong. He reached the other side without too much effort and signaled the unit to start over. At this point the river was 10-15 meters across and had a lot of cover on both sides. It was my turn to get into the stream and as I moved forward the water rose to my chest, but it felt good and cool against my skin. I held my weapon up over my head to keep it dry. It took the company about 30 minutes to get everyone over and assembled on the far side. The wet fatigues felt cool and refreshing against the skin, but they would dry out soon enough.

We continued on our sweep of the area, moving through the checkpoints. We reached the village, but today, we just skirted the outer edge and continued on. Around 1300 hrs we had reached our outer marker and began to circle back toward our NL. We were back into hedgerows. It was like a whole bunch of yards with fences, only the fences were made up of trees, bamboo, bushes, and grasses. Each hedgerow could be raised like a dyke. That was both good and bad as it offered cover for the VC to ambush us and it gave us cover to get behind if needed. Each hedgerow area could be several hundred feet square, to plots covering 1,2,3 or more acres. Within the hedgerows are grass fields, rice paddies, scattered fruit trees, banana trees, or just plain jungle.

The company was making good time with little to show for it until Charlie decided now would be a good time to ambush us like he always did. The VC just doesn't want to go toe to toe with us so they use their guerrilla warfare tactics on us. This time, they hit us with some AW fire and fired three RPG's at us. In the first volley, Ken Christianson and Steve Ward were hit. Ken took an AK-47 in his calf muscle and Ward got it in the leg and right shoulder. The guys up front in the column started shooting back in the direction of the fire. Hoping to out flank the gooks and catch them in a cross fire, I took my squad up the right side. We lay down a burst of fire to our left then started moving up the hedgerow, spraying the underbrush as we went. I told Wales to use his M79 to lay down a volley of fire along the tree line to our front and to our left to flush out the enemy. I had Jesse Tostado do the same to the hedgerow on the right. We got out in front of the company about 150 meters but found no VC willing to engage us. I ordered the squad back to the platoon.

In the meantime, the company had radioed for a Dustoff for our

WIA's. I wouldn't see Ward again until 2013. He told me that after he was flown to 12th Evac, he was transferred to Japan to recover. It was there that he said "I felt guilty about not being back with the company for the first few weeks. Then, it dawned on me the seriousness of the situation. About that time rumors were floating around that my platoon, 3rd, had been wiped out and now I didn't want to go back to Vietnam. After several months there, I was transferred stateside for further recuperation." Ken returned to the field later after being patched up. Our paths crossed again after Nam. Ken and I both ended up being stationed stateside at Fort Ord, California after my tour was over. Once the guys were airlifted to Cu Chi, we worked our way back to Stuart without any further incident. Two days later, the company was ordered to Cu Chi for a three day stand down. For our few days out of the field, we got to relax, get clean fatigues, and spend some time at the NCO or enlisted men's club and have a few beers and listen to the local Vietnamese bands play American Rock and Roll. It took our minds off of the war for a brief period. After the stand down, we were trucked back to Trang Bang by convoy. On the return trip, a 3/4 ton vehicle just in front of us hit a mine on the road and the explosion killed the passenger next to the driver. Our driver hit the brakes and we scrambled out of the deuce and a half and ran up to the vehicle. The driver was shaken up and his ears were bleeding. He asked how his buddy is. We found him lying by the side of the road with his entrails and stomach spread out in the dirt. We covered him up. The vehicle was totaled. It was another wait for graves registration and a tow for the vehicle. The rest of the convoy moved on. Once this mess was cleaned up, we departed as well and headed on to Trang Bang.

 I wrote home asking my mother to send me some fresh fruit. She was quite accommodating when it came down to it. Later, I got some apples, a little bruised up, and more cookies which I shared with the guys and made them feel like they were home. I also got some dried apricots to munch on, anything to get away from a steady diet of C-rations. I could only wolf down so much of that shit. My stomach was on fire and I couldn't shake the burning going on down there. It was probably from the tension and nerves, or maybe some of the bad food was contributing as well. A visit to Doc Sheppard got me another bottle of antacid tablets which helped squelch some of the nagging burning I felt in the pit of my stomach.

On October 23rd I told Wales to take the point. I used him whenever third platoon had the lead and my squad was assigned to act as guide for the operation. Sometimes we took turns on point. We took a few pot shots from Charlie today. Later Wales stepped on a punji stake in a trap but nothing serious and the stake just grazed him. For that, he got a Purple Heart. The medic dressed him up and we continued with our sweep with no further contact. We returned to FSB Stuart around 1530 hrs, just in time to get a nice cooling sprinkle before we entered the perimeter wire.

Wales was from Queens in New York and has a refined Queens accent if I could describe it as such. He wore a large medallion around his neck which some guys didn't look upon with favor. They thought it was a peace symbol but it was some kind of Chinese luck symbol. Wally shrugged that off and went about his business. He and I were really close. He was somewhat a joker and always had something to say which kept the squad loose.

Hill was also from New York, Astoria I think. He was a big strong black kid who was very laid back. He was easy to like and got along with everyone. Hill liked to wear sunglasses and donned a black embroidered shirt when he was trying to chill out. Price was from Loretto, Kentucky and the hill country. He was very much a country boy, somewhat slow to talk, with a little accent, not too much. Price was a gunner, who operated an M60 and he was very good at it. He could lay down a field of fire shooting from the hip and had a fine appreciation for the talent he had using his weapon. He was dependable and became another of my close friends, along with Dennis Buckley, whom I hung out a lot with, along with Hill, Tostado, Keatts and a few others. Dennis came from Burbank, California. We fellow Californians could really talk up the Beach Boy theme. I think I was the only dude from Northern California in the platoon other than Richard Coleman. I found out later he lived in Fairfield, about an hour from my house.

To this group, we only saw each other for what we were, soldiers. There was no color or race barrier for us. Everyone believed that we had each other's backs and would do whatever it took to get us all home alive. We believed this, but we knew some event would be out of our hands. Funny, but over the course of a year, I don't remember really talking about faith or religion with anyone. Maybe it's on everyone's minds so it just goes unspoken.

October 25th, Delta Company's 3rd platoon was out to the northeast of FSB Pershing. They were led by 1LT Edward Golder III who was new in country, having arrived in September. Golder's platoon, along with his company, were operating up near the Mushroom. If one were to look at a map of the area, the Mushroom was formed by the lazy tracing of the Saigon River as it winds its way down toward the capital city of the same name. As the company was sweeping the area, they made contact with a VC element. Rather than having another platoon execute a flanking movement to pinch the VC, Golder took it upon himself to lead his platoon on an assault into a tree line. A single shot rang out, missing the man to his front and striking Golder in the chest. As he was falling backward, he said, "I'm hit, I'm hit." He died when he hit the ground.

Rumors had a way of circulating around the company areas. No one knows who starts them and most are unfounded. We heard stories about individuals on occasion and although we didn't socialize with other platoons or companies, some names did become familiar to us. Jesse Anderson and Elmer Lightner from first platoon, were getting reassigned to *resupply* and join Sidney Fowler back in Cu Chi where the company HQ was now located. We also heard the exploitations of SGT Craig Schoonderwoerd, also from first platoon who was a calm leader in battle.

Our medics were great and there are no words that can show appreciation for what they did. It was amazing to discover how many medics we lost over there under fire. Most people I believe would have this idea that the medic had large red and white crosses on his helmet or some other form of recognition to indicate he was unarmed. That practice was stopped in Korea when the wearing of that red bullseye only encouraged the enemy to shoot at the medic. They didn't honor the Geneva Convention so it was decided to stop giving him one.

To Charlie, the medic was one of his favorite targets. He seemed to take extreme pleasure in shooting them. I know we had a bunch of them attached to our company. The medics listed here mostly served with Charlie Co during my tour, although you will find medics went where they were assigned or attached, many also went on patrols with other companies if needed. First there is Andy Wahrenbrock who was with 2nd platoon and was hit by our own gunships and came home early. He served only four months. Wade Lasister was followed by Jamie Ceballos, Dennis Sheppard, Jim Phares, Charlie Soule, Doc Lawrence, Terry Robinson, Clarence Olson, Rondual Tice, Leavy Solomon (KIA), Gary Green, Bob Meyers and later on, Leonard Dodson (KIA), William Kindle (KIA) and Peter Gerry (KIA). There are many more that served before and after I departed who all did a wonderful job.

September Arrivals: John Ausburn, Sand Springs, Oklahoma, Ellis Barnes, Detroit, Michigan, Richard Coleson, Ely, Iowa, Alex

Conley, Hampton, Virginia, Chester Novak, Oak Lawn, Illinois, Richard Satterswaithe, Butte, Montana, "Doc" Leavy Solomon, Palmetto, Georgia and Jesse Tostado from San Bernardino, California, a total of 25 men. On their way home was Kjell Solberg, whom everyone thought was crazy. He would ask if anyone wanted to sneak out and kill VC at night. No one wanted to go to that party. Also leaving was Platoon SGT Harold Steele Jr.

October Arrivals: Vernon Becker, Denver, Colorado, Donald McKenzie, Lumberton, North Carolina, Bruce Reed from Peekskill, New York and Robert Walton, a total of seven men for the company, of which three came to C Co. We lost another really good and experienced NCO in October, Hugh Bishop III. I'm happy for him and that he made it through his tour. I often wondered how he made out after the war.

17. AIR AMBULANCE — HOW ITS CALL SIGN CAME TO BE

As the U.S. military expanded its ground operations in Vietnam, so too was the need for supporting medical staff, a given fact in any war. Most of those who arrived in country in 1966 or somewhere on that time continuum, were the lucky recipients of individual efforts to narrow the time it took to provide on-sight medical care as well as shortening the time it took to arrive at lifesaving medical attachments strategically placed around Vietnam.

The medical arm of the military consists of field hospitals, surgical units, medical battalions, medical groups, evacuation hospitals, convalescent centers, AND Medical Detachments (air ambulances), a service dating back to the Korean War. The term MEDEVAC means medical evacuation of some type, and in the case of Vietnam, we generally think of the Huey helicopter air ambulance.

For those who were assigned to the 25th ID area of operation in III Corp, we were supported by the 12th Evacuation Hospital in Cu Chi. The 12th Evac (call sign Golden Empire) was also supported by the 60 bed 7th Surgical Hospital located next to it in Cu Chi which fed their wounded down to the 3rd Field Hospital in Tan Son

Nhut for more severe cases. There was also a 45th Surgical Hospital (MASH) located in Tay Ninh.

Whenever someone was wounded or killed, the RTO would call for a Dustoff. Many of us may have used the term Medevac as well. Was there a difference? The answer to that question lays in the initial deployment of the 57th Medical Detachment back in April, 1962. When they first arrived in country at Nha Trang, commanded by John Temperilli, they had no operational guidelines, doctrine, or support and had five Huey aircraft to cover the entire country. Initially, they were instructed to only provide medical aid to U.S. troops in spite of the fact that it was the Vietnamese that needed the help with U.S. casualties only amounting to 30 individuals at that time.

In 1963, the unit was moved to Saigon and Major Charles Kelly became the unit's third commander. The unit's call sign was *DUST-OFF* so taken by Lloyd Spencer, the 2nd commander, and was approved by the Navy Support Activity, which controlled call signs. It was fitting because some choppers disappeared in a cloud of dust when landing during the dry season. It caught on and because the singular call for medical assistance. The call sign also generated its own radio frequency which eased the lines of communication between the ground units and the medical detachment.

It was Major Charles Kelly, and two pilots who joined the unit about the same time, Dick Anderson and Patrick Brady, who developed their operational standards for performance. Those standards were to respond without hesitation to all calls, to be there as quickly as possible, and to service whoever needed aid, friendly and foe. Higher command frowned upon the 57th operational guidelines, and to be exact General Stillwell who was Kelly's boss, wanted operational control of the unit in spite of having no knowledge what the 57th truly did and how they needed to effectively operate. This created tremendous tension between the General and the unit. Eventually cooler heads prevailed and the unit was left alone, but it remained controversial because of the risks they continued to take with both day and night time operations.

The unit was moved once more to Soc Trang where there was only ARVN air support. The unit was quickly reinforced by the newly arriving 82nd Medical Detachment. The men of the 57th volunteered to go to the 82nd to help train them in their methodology. *Kelly's Krazies*, as they were known, were highly loyal to their commander and to the unit. They were truly respected by the men on the ground and knew if they contacted the 57th, their boys would be carried to safety regardless of the situation. Most of the transfers were approved but this did not help the new commander of the 82nd who did not believe that the Kelly way was the answer and demanded that he personally approve every mission. This led to a lot of angst among the pilots.

Soon after, there was a debate as to what call sign the 82nd should use. Some in the 57th did not want this unit to use their call sign *Dustoff* because they did not want the tradition of Dustoff to be diminished by some who disagreed with the operational principals of the 57th. The argument was lost and consequently every new medical detachment to Vietnam used Dustoff for their call sign, except for those units that operated in the 1st Air Calvary. They would be known as MEDEVAC for they were different than the rest of the medical detachments. Some of the differences were not responding to the aid of ARVN or local villagers, hot LZ's, night flying, or weather challenging missions. It was clear to most that Dustoff units had a higher standard than Medevac units.

So, in spite of the fact as an RTO myself, we found the usage of the term for medical assistance to be either Dustoff or needing a

Medevac as they were interchangeable. When we were on the radio on their push (radio frequency), it was "Dustoff 159 this is Charlie 36 X-ray, do you copy?" In 1967, the 159th Medical Detachment arrived in Cu Chi and served as our Dustoff until we left Vietnam.

A great book to read is called *Dead Men Flying* by Patrick Brady that details the story of the 57th Medical Detachment and is the basis for this chapter.

18. FIRE SUPPORT BASE PERSHING

NOVEMBER 1968

The rest of the battalion had been in the area of what would become FSB Pershing for about two weeks when they started to set up the FSB on October 3rd while we were laboring away down at FSB Stuart. Several weeks later, Charlie Co arrived and joined the battalion.

At the FSB site the ground was concrete hard and our small portable G.I. shovels couldn't make a dent in the soil. Most of the bunkers had been built but some needed improvements and we needed some trenches built between the firing positions. We got permission to use explosives to speed up our work. With my training in demolitions, this would be a cinch. I got several hundred feet of detonation cord or *det cord* for short. That is a white opaque thin plastic rope filled with PETN, pentrite, an explosive that burns at a rate of 6,400 ft per millisecond. Every three to four feet, I made a knot in the cord. Once I had done that, I cut a piece of fuse, about six inches in length, and attached a blasting cap to it and then taped the blasting cap to the end of the det cord. Next, the cord was laid on the ground along the path in which we wanted to create a trench.

After giving the traditional yell of *"Fire in the hole!"* I lit the fuse and stepped back. The det cord was very powerful and was used to link charges together. Because it burns so fast, it allowed charges to detonate at the same time when they are daisy chained together. It acts like an explosive charge, but is really a fuse. The impact of the det cord exploding caused enough force to loosen the soil down several feet. It makes for easy shovel work and then we can repeat the process.

The concertina wire that was strung around the FSB was addressed next by adding empty cans with rocks to rattle when moved and a series of Claymore mines were set up along with Foo Gas, which is jellied gasoline, and other nasty devices that can be detonated to break the back of the enemy should he choose to mount a ground assault. One of the entrances to the FSB through the wire was directly in front of our platoon's defensive positions. The entrance was a winding snake like affair that took time to walk through. It was not a straight shot into the FSB. Why? In case the enemy got into one of the entrances, it would take time to walk the zig zag built entrance, thus giving the defenders time to eliminate the threat.

The FSB was equipped with a battery of 105mm howitzers and 4.2 inch mortars. We watched as the artillery conducted a firing test. A previously little-used 105mm howitzer technique of firing a short-time-fused projectile at a low quadrant with *charge one* had proved very effective. Data was computed from the TFT (tabular firing table) for a 5-20 meter height of burst at ranges of 200—1000 meters from the weapon.

This technique was pioneered for the 25th DIVARTY by the 1st Bn, 8th Artillery while commanded by LTC Robert A. Dean (in the battle of FSB Burt, vicinity XT4908, on 1-2 January 1968). The theory and practice of the technique were further developed between January and May. On 12th of May it was again successfully employed at the battle of FSB Pike VI vicinity XS7395. Since then, the method had been used extensively with great success indicated by secondary explosions and high enemy body counts after all subsequent ground attacks on fire support bases. The technique was known locally as *Killer Junior*, a name derived from the radio call sign, *Killer* (by which 1/8 Arty was formerly known). Killer Junior provided effective interdiction while the enemy was attempting to mass for an attack, and during the attack would annihilate his assault formations. This method used HE (high explosive) and was different from the beehive or flechette cartridge rounds which were filled with darts and used as an anti-personnel defense.

I was getting excited about the prospect of going on R&R. I had gotten travel orders for Australia and I would be leaving within the week. For the next month and a half our battalion was split between FSB Stuart and Pershing. Every day one or all companies were making air assaults to the area to our east and north. With each excursion we were taking on hostile fire and our causalities were mounting.

November 6th, we were ordered to saddle up. To the south and east of us, we were needed to provide security for A Troop 3/4th Cav. They had two tanks stuck up to the turret about a mile off the main road. One tank was not operational after throwing its track trying to extract the first one. Our platoon got the job of providing added firepower and security for the night.

The next morning we retraced our steps to the main road where we met up with two M88 Tank Retrievers sent out from Cu Chi and escorted the tracks back to A Troop. The TR is a monster. Its tracks appear to be about two feet wide and it carries a large boom along with a winch designed to recover damaged tanks. The engine is a Rolls Royce that spits a blue flame when running. We were all wondering if this track could get close enough to the stuck tanks to help out or would it become a casualty to the mud too.

As we arrived at the site, we had attracted the attention of the VC.

We started receiving sniper fire from a tree line about 200 meters to the north and returned fire. I was getting annoyed and fired a LAW (light anti-tank weapon) toward the tree line, sending a message to the VC to lay off. After a half hour of this, Charlie felt he had done his job and retreated into the jungle. The crews from A Troop and the TR started back to work hooking up cables and using each other as dead weights in order to suck the M48 out of the mud. It took several hours to get the job done. However, no sooner had we gotten one unit unstuck and we had to pull another tank out. We played tag all day, either extracting a tank or spending time watching the crews wrestle to get one of their tracks back on the road wheels.

By late that afternoon, APC's, tanks, and the TR's were parked on the highway. We were all tired and took a break before we set out to get back to our NL. The M88's headed back to Cu Chi with an escort, and we got a ride back to FSB Stuart with the tracks. Sitting on a tank is a thrill. Charlie won't fuc* with this big boy, not today anyway. No one had been hurt in the operation. It had been a good day. The boys were looking forward to a hot meal and some mail from home. Today was goody pack day too and most of us were out of cigarettes.

On the 9th of November, my R&R (rest and relaxation) started. Basically, it was a brief vacation, of which we could choose to go to Bangkok, Singapore, Tokyo, Manila, Hawaii, Sydney, Hong Kong, Kuala Lampur. I chose to go to Sydney, besides traveling to Sidney, it was a six day trip while the others were five day excursions. The flight to Sydney took over nine hours with a stop in Darwin where I tried drinking a Foster's beer that was twice as large as an American beer. It must have been 5 or 6% alcohol because I felt pretty good after that one. We landed at 0600 at the airport. The first stop was to find a hotel and I picked one out called the Whitehall somewhere in town and took a taxi there. As I walked through the lobby, I could hear the Beatles playing over the speakers singing *Hey Jude, don't make it bad, take a sad song and make it better, remember to let her into your heart, then you can start to make it better.* I shared a room with another soldier whom I met on the plane on the way down there. I had a great time visiting down under. The people were all excited to see Americans, especially those in uniforms. They went out of their way to make you feel at home.

The hot spot was at the Crossroads downtown. There, you could find plenty of saloons or bars and go-go clubs to frequent. The women were plentiful and quite willing to share their apartments with the GI's if you were fortunate to find one who took a liking to you. It was not uncommon to have some gal pull up in a taxi and ask you if you wanted a good time for $10. Anyway, it was an experience I will never forget. When I first arrived there, I needed to get out of my khaki uniform so I went shopping for clothes and found a nice shop where I picked out a pair of slacks, shirt, and sweater. I needed shoes and the man in the store said he had just the answer. He took me by the arm and out the door we went, the back door. We made a quick left and started to walk down an alley. About half way down the alley toward the other side of the block I saw another man coming my way and I began to think I had been set up, But, it turned out this man was the brother of the guy I was with and he owned a shoe store. So I went to his store, and got a nice pair of shoes to wear with the rest of my outfit. I couldn't walk down the street without the people stopping me and wanting to talk. I don't know how they knew I was an American, but nonetheless, it happened.

I took one day to go deep sea fishing about twenty miles up the coast from Sydney and didn't have any luck. It was more about not getting sea sick than the fishing, but I managed to get through the day. At the time I was there, the famous opera house that sits at the edge of Sydney Bay, was just being finished. I had a few photos of myself standing next to the structure. It was an impressive site to see.

The R&R ended quickly and I found myself back on a plane heading to Vietnam. We landed back at Tan Son Nhut and those returning from R&R usually stayed in a barracks for the night. Flight arrangements had to be made to get back to the unit and you usually could not pull that off landing late in the day or later at night. I left my things there in the barracks to get something to eat. When I came back I found out that I had been robbed. That was a real downer having my camera and film stolen while I went to dinner after I had landed in Saigon. I couldn't even trust some fellow G.I.'s to watch my belongings. I was back with the unit in no time sharing my trip. The guys that seemed to have the most fun went to either Bangkok or to Manila. The married guys all went to Hawaii to see their wives.

Long gone was the monsoon season but we did get some rainfall with an afternoon shower that washed away the sweat of the day. The problem, if you wanted to call it one, was that it would always rain when you were about a 1/2 hour away from the FSB. The timing was such that wet clothes were always the attire that you just had to deal with until they dried out. The jungle fatigues we wore were great in the sense that they held up well, and did dry out reasonably quick.

Our steel helmets or pots, caps, lids, take your pick for names, were supposed to be used as a head cover for protection. From what, I'm not exactly sure. Maybe they kept us from bumping our heads against hard objects like the steel plate used in some of our bunkers, or banging our heads as we entered or exited a Huey. But as far as keeping us safe from bullets and shrapnel, I wouldn't bet my life on it. Most of the guys had written names of their girlfriends, wives, states, hometowns, or cuss words somewhere on the camouflage cloth cover that is held in place by a large elastic band over the steel pot. Mine looks like the rest of them, covered in names with a slogan on the band that reads "California My Way."

Watching movies and reading stories about WWII, the steel pot seemed to be used for other things as well. We had seen it used to hold water for shaving and washing. Those tasks seemed to be one of the normal activities observed. The other one, from what I have read, was not so good. What one? Cooking food in your pot? The material (steel or some blend of metals?) used to make the pot apparently was weakened if heated. There was only one time I remember while in Vietnam, and that was at FSB Pershing, that I saw a steel pot used in that manner. I was part of it. Late one night after skipping or not liking what was served at the mess tent for dinner, a few of us decided we wanted a late evening snack. So, what was a grunt to do? We tried to raid the mess tent. Well, we actually did raid the mess tent, looking for something edible. That can be a challenge when you're in the field and dealing with army cooks. The only item we found were chicken eggs so we took a dozen or so back to our company area.

This wasn't far away since we were located almost next door to the mess tent anyway. Upon returning to our home turf, we got some water going in someone's pot and heated it up with the C4 we carried with us for just this type of situation. Even boiling the eggs couldn't rescue them from the taste of being old and not fresh. We tried to find something to cut into the flavor, mayo, mustard, and sweet pickle, but that didn't do it either. After that experience word got out somehow that someone had raided the mess tent and asses would be had if they could find the culprits. We spread the rumor that we thought it was guys over in Delta Company.

Our home at Pershing was not without its drama. We received mortar fire at least once a week since we had taken up residency. Sometimes it was just a single round from an 82mm. I think this was Charlie's way of just having a laugh at our expense. Other times, he may decide to give us 8-10 rounds. It always varied and never had any pattern to it. When you heard that distant *bloop, bloop, bloop*, it was customary to start singing *incoming* and then start looking for shelter.

Our gunners in the 81mm pits and our 4.2 inch mortar platoons were pretty damn good. One night when we were getting some incoming fire, they jumped on their tubes and returned the fire. They actually got a secondary explosion from the woods. Maybe they got the enemy gunner that night.

In the middle of a war, some people found objects to hold on to that had some meaning to life. For Joe Liberator back in Dau Tieng, it was adopting a puppy. For others, it took a spider monkey or two. At Pershing, in our company area we had a grove of bamboo plus a few trees, the only vegetation in the entire FSB. The platoon had a pet monkey or the monkey had a few owners, one being Dennis Buckley and the other was Richard Coleson, our RTO at the time. The monkey was named Ho Chi Minh and Ho did not care for Bob Walton, at all. Bob's nickname was *Killer*, although there was no intimidating look to the guy.

Now, Ho Chi Minh was usually on a chain near the bamboo. The guys would take turns packing Ho around and feeding him. I did not get involved with these monkeys as I had no desire to get scratched or bit. I'm sure Ho would have preferred to be free and roaming the jungle. One day Ho got lose somehow, and spotted *Killer* Walton and took out after him. It was quite a scene seeing Walton fleeing for his life as he disappeared to the other side of the FSB. Eventually Ho escaped his captivity.

There was another problem we had when we were rotated between FSB Stuart and Pershing. The locals were well familiar with pay day. About a day before we got paid, the prostitutes or Boom-Boom girls as we called them, would start hanging around the bridge at Stuart. There would be one or two girls and they would set up a makeshift tent on the side of the road and wait for customers. Someone would walk by and you could hear, "Love me long time G.I.? Only $3." The customers they got for their $3 service, would visit until their money ran out, then the ladies of the night would move on until next payday.

The Boom-Boom girls were never very far away as it was, they were always close by. The officers never said anything about this, nor did I. Not being a customer, I just looked the other way as we had bigger problems than that. Unfortunately, the command did monitor our health as a battalion, and they kept track of VD cases. Usually a week later after payday, the First Aid station would get overwhelmed with customers. CPT Charles Boyd would recall later a staff meeting he attended. Charles was C.O. of A Co beginning in February, 1969. The good news at the staff meeting was A Co was on top of the battalion in enemy kills, and the bad news was he was also the leader in VD cases.

During the month of November, I spent more time flying in choppers than any other month I was there. Our company flew 16 combat assaults into the Citadel, near the Filhol Rubber Plantation, Hobo Woods, small villages and hamlets trying to disrupt the supply routes of Charlie. This was the 271st, 272nd, and 88th NVA regiment area. We had a lot of sporadic contact with the enemy but we did not encounter any large forces. As luck would have it, it seemed we managed to be avoided by them. Alpha, Bravo, and Delta companies had the heavier action on more than one occasion. Several times they sustained ground night attacks but managed to repel the enemy. Some of these fights involved hand to hand combat.

It was about this time in mid-November that division had decided that we should receive our resupplies by truck rather than airlifted in. This meant that every morning the 65th Combat engineers left Cu Chi which was down the road about 10 miles, and began to sweep Hwy 1 for mines. This proved to be an effort in futility because the mines the VC set had very little metal in them. They used diesel fuel and ammonia nitrate along with a hunk of bamboo, a nail, and a bullet to act as a detonator. They would bury this recipe in the road and wait for a truck or tank to run over it. The explosion usually destroyed the equipment and occasionally would kill the occupants.

As the engineers started their walk from Cu Chi, we would escort another group from FSB Pershing to sweep down road TL-6A toward Trang Bang. This dirt road tied into the MRS, main supply route, known as QL1 or Hwy 1. Each morning one of the platoons from the base would be assigned escort and patrol duty until the convoy arrived, unloaded and returned to Cu Chi.

As a traveler passing through Trang Bang to reach the junction of road 6 Alpha as we called it, you would travel through the market place. This was a very busy place as residents, farmers and shop owners exchanged words and currency for goods purchased, Eventually, the shops turned into residential housing which gradually thinned out. Near the edge of the village and just to the side of the road, stood an ARVN compound. This fortified structure was surrounded with concertina wire, Claymore mines, and the building was protected with sandbags and gun ports. There was a local militia stationed there of populist forces where upon occasion, joined the U.S. Army in fighting the VC. We usually chuckled at this loosely run group of military personal. This war seemed more like a nuisance to them, than it caused them to really grab their full attention. The compound was one of our checkpoints we used when on road patrol.

The problem we faced with the daily patrol was being predictable to the VC. Soon they started taking potshots at us and then they began to set booby traps along the edges of the road where we walked. Some of the area passed through a cemetery and then there was a stretch of bombed out houses with tons of rubble. This was easy going for the VC to place their deadly traps. I found myself stopping for a rest one time and after I had sat down on an abandoned chair, I noticed a trip wire running right under where I was sitting. After carefully getting up, I traced out the wire and defused the grenade.

Among the helicopters that we saw coming and going, was the Hughes OH-6A Cayuse, used mostly by the Battalion commanders.

The Brigade and Division officers preferred to use the Huey Slicks, primarily for space needs. The amount of radios that were packed into the back of the chopper in the passenger area took a ton of space. All those frequencies that the command needed to monitor or contact was vital to the C&C structure (command and control). Back to the battalion officers and their use of the Cayuse, a very nimble and fast moving chopper, they reminded me of a large bee. There were several versions used, mostly the OH-6A, but there was a C model and it set itself apart by having five rotor blades.

The chopper nicknamed the *Loach* after the acronym LOH — Light Observation Helicopter, could hold five people; any more than that and the Cayuse couldn't lift off. Not much good, matter of fact, no good for close in support. One afternoon, late around 1500, I got a call to see the Lieutenant. HQ wanted to check out an area east of FSB Pershing on the ground and use our platoon to do it. The LT and I would fly a reconnaissance mission and check out the area before we jumped off the next day.

Out on the helipad was an OH-6A parked and waiting. After our briefing, the LT and I walked on out to the pad, and climbed aboard. The pilot started the rotors turning and warmed up the engine. A minute later, we lifted off and headed east to our coordinates. Now, normally one would think, let's not alert the enemy we are coming. I guess we tried to be somewhat inconspicuous in the process as we buzzed over the tree tops and swept over hamlets and rice paddies on our way to and from our destination. It was a blast riding in that Cayuse. For a brief moment, I had a chance to remove myself from the aspects of war while riding up there in the air. I felt free and removed from it all. The flight seemed all too short for me. Forty-five minutes later we were setting back down on the pad at the base. The flight and journey were over. We got the intelligence we needed for the next day. That was my only ride in the Cayuse, but I still think about that day whenever I see a photo of one. The next day, 3rd Plt flew out to the area we recon'd the previous day and searched the area, not exactly sure what we were expected to find there. It turned out to be a quiet day, no contact and we left empty handed back to Pershing.

Life at Pershing was fairly quiet, primarily because of its enormous size, housing a full battalion of infantry plus a battery of 105mm, Four-Deuce, and 81mm mortar support. Last night, the Four-Deuce boys (4.2 inch mortar) had a fire mission. The four deuce mortar was the only piece of hardware, other than the 8" howitzer, that was referred to in inches and not millimeters when defining the size of the firing tube or muzzle in the Army. Both are now out of the active army arsenal. The 4.2 inch mortar replacement was the 120mm mortar.

It was dark and late into the evening when we could hear someone *in contact* out to our southeast. We did not know if this was an ambush patrol that was from our base or possibly from FSB Stuart, but nonetheless, someone needed artillery support. The Four-Deuce was good for a maximum range of four clicks or about 2.5 miles using HE.

As we were sitting by our bunkers about to wrap up the evening conversation and get some sleep, the Four-Deuce started their fire mission. Soon, we heard the loud *BLOOP, BLOOP, BLOOP* of the rounds leaving the tubes. Quickly after the initial volley, as the second firing started, we hear "*SHORT ROUND*", then more firing, then again "*SHORT ROUND*". We lost count, but every time this phrase was blasted out, we took cover, because the projectile was not flying out of the mortar tube, but wobbling out, meaning the powder charge was no good, maybe wet. The good news is that the projectile must travel several hundred meters before the counterweights in the fuse rotate out away from the firing pin arming the round. These rounds were falling harmlessly outside our concertina wire. The next morning, my platoon was sent outside the wire to retrieve these 25 lb rounds and bring them back to the Four-Deuce boys.

Near the end of the month, our PL, 1LT David Riggs, was moved up to battalion and replaced by 1LT James Merrett. Talking to him, he appears to be a nice guy. He's from Phoenix, Arizona. This was his first field command. He just got engaged to be married before he came overseas. We spent some evenings chit-chatting with him, trying to size him up. He was soft-spoken and subdued in tone. The NCO's did our best to give him a tactical summary of the platoon's capabilities and what we are facing in our area of operation.

Some of the guys were getting bored that evening waiting for the sun to retreat down below the horizon. Most of us carried a bayonet and soon we were playing *Mumbely Peg*. I guess there are numerous versions of this game. The one we played involved two players who stood face to face and about three to four feet apart. The first person would attempt to stick his knife in the ground next to his opponent's foot. If he was successful, the opponent would slide his foot next to the knife and keep it there. He then made the same attempt in return against his opponent, the idea was to not stick your opponent's foot but to cause your opponent to continue to widen his stance until he could no longer stand or reach the stuck knife or make an attempt at his own to stick a knife next to his opponent's foot. There were plenty of holes in boot tops, but no one tried to stop this game from what I could remember. It would not surprise me however, if someone ended up seeing a medic, they would be in jeopardy of getting an Article 15 if it meant having to go on light duty due to injury.

There were days when getting any break from combat duties was a great distraction. When the resupply convoy rolled in, it was a chance to shoot the shit with some of the guys who left the field combat for various reasons, or sometimes, just luck, that we don't get to see often. Some of these guys were fellow NCO's so the latest scuttlebutt could be juicy news. I spent an hour with Sidney Fowler, Jessie Anderson, Elmer Lightner, Hiram Marziano, and Greg Nevers. Except for Sidney, they were all from 1st Plt. They all arrived at the same time as me, so we kept close tabs on how everyone was doing. Hiram mostly stayed in Cu Chi, but today he had some important documents to deliver to Battalion. They took turns driving the "point" Jeep and taking up the rear guard of the column eating everyone else's dust. Wearing goggles was necessary when they were coming down Six Alpha. On those trips, Greg Nevers, who was an RTO for a while, always seemed to find his way into the various photo shoots that someone in the platoon was orchestrating. I was happy to see these guys and wished them a safe trip back to division basecamp as I headed back to my platoon area.

We received visits from dignitaries on occasion too. Senator Ralph Yarbourgh of Texas and four-star general James K Woolnough came through to see what a war looks like. Woolnaugh was in charge of all U.S. Continental forces in the United States. We also spent a delightful day being entertained by a group of American Red Cross workers, i.e. Donut Dollies out from Cu Chi. They were all decked out in their blue uniforms and it was a pleasure just seeing an American girl up close.

New replacements for November: 1LT James Merrett and 1LT Michael Sheehan.

19. DECEMBER, OUR WORST MONTH

In between the mine sweeping security and road patrols was an occasional combat air assault or RIF around the basecamp. It was December 3rd and on one of these operations we were at two company strength, B and C companies. After landing at the LZ, we got into tactical formation with B Co taking the point. After about an hour of walking, the patrol pulled up to rest as the lead platoon from B Company stopped and was checking something out. C Co, following, had our flankers out as usual. Jimmy Sheffield in 1st Platoon was on flank and tripped a booby trapped 8" howitzer round. There was a *huge* explosion that rocked us and the surrounding area. There were shell fragments whizzing through the air and striking the ground all around us. One hit just to my right. As a dumb ass, I tried to pick it up and burned my fingertips. Just looking at the jagged, chisel sharp edges to that fragment gave me the willies. At the time, my squad had happened to hit the deck to rest our feet and grab a smoke. That shell killed five men and wounded another eight that required several Dustoffs. It was a tremendous explosion that caught us all by surprise and the concussion was felt by everyone. This was a sad day for us. We lost John Ausbern, Don Bonner, Alex Conley, Jimmy Sheffield, and Arthur Warner in that explosion.

It didn't take long for battalion to dream up some new tactics to roll out. At first we just patrolled out to Trang Bang and back with two squads. Sometime in the second week of December, it was decided that we would leave a small outpost at the edge of Trang Bang. It worked this way. We would start out with a platoon size patrol. Escort the engineers down to Trang Bang using two squads where we would hand them off to another platoon coming from FSB Stuart. At that point, we would drop off three troops at Trang Bang, then the remainder of the squad would patrol back to the midpoint checkpoint along route 6A, where we would meet up with our other squad which looped from the midpoint back to the intersection of 6 Alpha and the entrance road to Pershing. We did this circular pattern until the convoy left Pershing and returned to Cu Chi. That way, the VC couldn't mine the road again. We were covering the road from both ends. Aside from the occasional sniper fire, we did ok.

December 10th was a busy day. All three companies were engaged in some type of action. First, Delta Co set off a booby trap and evacs two wounded via Dustoff, then later that morning engaged a squad of VC with small arms, artillery, and air support. At 1220 Bravo Co was being fired at by two or three VC. They engaged with organic (means using what was available and belonging to the unit) weapons and had one WIA. After lunch, Charlie Co split up and 1LT Louis "Duke" Gieka's 1st platoon along with 1LT Hawthorne's 2nd continued on with a RIF while 1LT David Rigg's 3rd Platoon was ordered off on another azimuth which would take it back to Pershing.

Around 1310 hrs, Charlie Co (-36) had been conducting a RIF toward a village. The company started receiving small arms and automatic weapons fire from an unknown number of VC. Fire was returned with small arms and 1LT Roger McDaniel called for artillery and gunship support from Diamondhead, 25th Avn Bde, Cu Chi.

The gunships arrived on station and began to attack the VC. They had a number of wounded and radioed for a Dustoff.

The company continued to engage the enemy and the fighting intensified. The Dustoff chopper finally arrived and the LZ was hot. The pilot attempted to set down and his aircraft was riddled with bullets so he pulled out to wait for a safer time to land. In the melee that pursued, SP4 William A. Lowry, the C.O.'s RTO was killed and McDaniel was hit in the chest with the same bullet that got Lowry. SSG Donald F. McKenzie was also killed outright. The C.O. continued to direct the fight but was growing weak from loss of blood. It had been several hours now and the situation was growing critical.

Diamondhead was monitoring the net and knew what the situation was on the ground. The injured men needed to be extracted and taken to the 12th Evac in Cu Chi. SP4 Ron Leonard, a door gunner on the gun ship talked to his pilot. He told him that we just couldn't leave them there without doing something. He talked the CWO (Chief Warrant Officer) into going down to pick up the wounded. The fire was intense and Ron was providing covering fire with his M60. The pilot got the aircraft on the ground. His ship was being riddled with bullets from the insurgents. SP4 Leonard kept on shooting suppressing fire until the wounded were loaded. His barrel was so hot from firing that it had melted and could no longer provide effective fire. The chopper lifted off with its precious cargo and headed to the hospital. Soon after that, the enemy broke off the engagement, knowing they had done their job. The company had lost its C.O., two KIA's and five WIA's. It was not known what damage we had inflicted on the enemy.

By now it was close to evening. Battalion decided to leave the unit in the field and the company set up a night defensive position and was reinforced by two platoons from Bravo Co, led by CPT Alan Wissinger, and hunkered down for the night.

When SP4 Leonard finally stepped on safe turf back at Cu Chi after his mission, he found five bullet holes through his fatigues, but miraculously was not hit. For Ron Leonard's bravery, he was awarded the Distinguished Flying Cross. CWO Fred Panhorst, who was the other supporting gunship pilot, call sign Diamondhead 21, flew his aircraft to draw fire while Leonard's ship swooped in for the pickup. For his part in the Bo Heo action, he was awarded the Silver Star.

This is from LTC Richard Wiggins, Ret. A Co C.O., who wrote to me about this day on January 1, 2010. "Roger W. McDaniel, we all called him Bud. He had been wounded a day or two before he was shot during a mortar attack at Pershing. The day he was shot, the X.O. had the company out and you all ran into a hornet's nest. The bullet that killed his RTO is the one that hit him, going in his chest and out the back. His spinal cord wasn't severed, but the shock of the bullet is what paralyzed him from the chest down. Once he was stabilized in country, he went to Japan, then back to the States, winding up at a VA hospital in Richmond, VA that specialized in those types of injuries. I think it was in Dec when he was shot, and he actually shouldn't have gone out but he wanted to be with the unit so the Bn Cmdr let him go. After he was wounded, they couldn't get a Medevac in because of all that was going on. I was monitoring him on the command net and it was like watching a movie in that his voice was strong and after several hours just kept getting weaker and weaker. They finally got him out late that day or early evening. I didn't see him again until we visited him at his home in Richmond."

December 11th arrived. It was early in the morning, around 0443, at the NL at XT545232. There were two platoons of Charlie Co (-36) and two from Bravo Co. The VC hit us with a barrage of mortar fire and attempted to breach our west perimeter to overwhelm us. We returned fire with everything we had. The NL had been dialed in for artillery support from FSB Stuart which had 105mm and 8 inch gun batteries. They immediately began to pepper the area surrounding our line of defense. Spectre, the C-130 gun ship was called in for support. From FSB Pershing, we could see the ribbons of red tracers coming down from the aircraft as they engaged the enemy with three Gatling guns, each putting out 3,000 rounds a minute. Every fifth round was a tracer round and with those weapons firing so fast it looked like you were holding a garden hose spraying red water. At Pershing, Buckley, Price, and I were trying to get information about what was happening to our buddies in the company and wondering why we weren't sent back out to reinforce the other two platoons. The fire from the gunships was too much for the enemy and they broke off the attack. Bravo Co lost two men, PFC Alfred J. Critelli and SP4 William J. Sugden and there were six WIA.'s. The results for the VC was not good. They had 16 killed and two POW's.

Article from the *Tropic Lightning Newspaper* dated February 3, 1969

WARRIORS, AIR STRIKES DOOM 16 VIET CONG

By SP4 Charles Haughey (A Co)

Second Brigade soldiers of the 2nd Battalion, 12th Infantry, killed 16 Viet Cong in two days of fierce fighting nine miles north of Cu Chi. The First platoon of the battalion's Charlie Company spotted oversized rice shocks near a distant weed line just outside the small jungle village of Bo Hea.

"The shocks were fat and looked as though they were hiding something," said platoon leader 1LT Louis M Gietka from Clarks Summit, Pa. "It seemed too obvious," he continued, "So I directed my grenadiers to drop M79 rounds into the area."

Just as the round landed to expose what appeared to be a mortar tube, the Fire Brigade soldiers began receiving .51 caliber fire from their right flank. Reacting quickly, Gietka maneuvered his platoon into a nearby hedgerow. By this time the Warriors were receiving fire from several directions. Air strikes, artillery, and Cobra gunships were called in to assist the battling infantrymen. As the gunships and Air Force jets took their toll, the enemy firepower slowed and finally stopped. The Charlie Company soldiers assaulted the enemy positions, overrunning some of the bunkers. *"We had moved about 200 meters when all hell broke loose,"* Gietka continued. *"We suddenly were being hit from three sides."* Maneuvering swiftly to the southwest, the Warriors for a short time lost contact with the estimated reinforced enemy company. They took this opportunity to form a night defensive position.

Roaring into a nearby clearing, helicopters landed five ships at a time depositing the Warrior's Bravo Company in the ravaged area, amid sporadic bursts of enemy small arms fire. Bravo Company, commanded by Captain Alan R. Wissinger, San Diego, joined the battle to reinforce the night laager. Working feverishly by the flickering light of aerial flares, the two companies prepared for a night battle. At 0430, the enemy began an all-out attack consisting of barrages of over 100 RPG rounds from all sides.

"The enemy tried to overrun our west defenses, but our machine-gunners stopped him before he really got started," said platoon leader Gietka who was himself wounded in action.

SGT Craig Schoonderwoerd, C Co, 1st Plt, received the Silver Star for taking out several machine gun positions using grenades while rallying his platoon to hold their positions and their side of the perimeter and rebuff the VC ground attack.

"The enemy started hitting our side of the perimeter. The MG crews were throwing out lead at an unbelievable rate. I told them to wait until they had a clear target and to stop wasting lead," Schoonderwoerd said. "There were several enemy machine guns setting up to our front and I grabbed some grenades and took them out before they inflicted damage, then moved from firing position to firing position bringing ammo and verbal support to my guys," he went on. "The boys on the Bravo side had their hands full too. I think the VC were trying to con the Bravo guys to come over to our side to help so they could bull rush that side of the perimeter, but it didn't work."

The morning of the 11th the brigade swiftly reacted to the morning attack and sent CPT Richard Wiggins's A Co 2/12th along with D Co 2/27th Wolfhounds and A Co 2/14th Golden Dragons into the area and engaged the enemy in another firefight that lasted several hours. A Co 2/12th loses were four men KIA and Bravo Co had two KIA's. By evening all the companies converged and set up a defensive position for the night.

On December 12th, C Co, 3rd Plt was pulling road patrol along 6 Alpha. We were walking parallel to the edges of the road between FSB Pershing and Trang Bang staying about 30-60 meters away. There were two six man patrols and a small three man force left at the far end. Each patrol was looping around a center check point (CP), with one patrol going to the far end near Trang Bang and the other looping back to a CP toward FSB Pershing. The far CP was near the ARVN compound and downtown Trang Bang, where we dropped off a team to anchor the west end of our patrol. Ron Stepsie, one of the machine gunners was not feeling well, and Jesse Tostado from my squad volunteered to take his duty walking patrol. 1LT James Merrett, gunner SP4 Ron Stepsie, and SP4 Robert Beltran remained in Trang Bang while we headed back to the center CP. They never came back alive. As they rested, they were jumped by a number of VC who came at them from a small trail leading back to a berm line near their location. Before they could react to the threat, all three of them were killed. Beltran was from Fremont, California, Stepsie was from Pennsylvania, and 1LT Merrett, as I mentioned earlier, was from Arizona.

That was a shock to the whole platoon, hearing about these guys. Tostado really took this hard, knowing he could have been killed rather than Stepsie. This would haunt him long after he returned home, dealing with survivor remorse. The guys were well liked and did their job. It was really quiet in camp for the next few days as everyone digested what had happened.

Having those boys killed just made us more determined not to get caught with our guard down. We heard that they were overrun by a large force of VC and hardly had a chance to fire their weapons. One thing you don't want to do is try and out duel an RPG.

The VC used these weapons very effectively against personnel. They were an anti-tank weapon, but put out some serious shrapnel and we had sustained quite a few casualties due to these weapons. With the loss of Merrett, SFC Partee was given the assignment of temporary platoon leader. I was promoted to acting platoon sergeant. I left the squad to Russell *"Zoom"* Zimmerman as the acting squad leader, although I had my reservations about his leadership and his off duty attraction to smoking pot.

On December 13th, Charlie Co was OPCON'd to A Troop 3/4 Cav. That attachment lasted until January 2nd. We were moved over to FSB Stuart. A Troop had a platoon of M-48 tanks which were used to patrol the roads between Go Dau Ha and Cu Chi. We spent the day on bridge security.

December 14th, we were making a company combat assault to the northeast of us. We landed in a big rice paddy bracketed by hedge rows several hundred meters to either side. It was a two platoon operation, about 60 men. Right after landing, the company formed up and 3rd Platoon had the point for the company with 2nd squad leading the operation headed up by *Big Jim* McInvale with SGT Richard

Conlin right behind him. As the column neared the hedgerow area, we started taking AW and machine gun fire. The company was pinned down out in the open.

We had landed in a rice paddy area that afforded us some protection because the dyke walls were high enough to shield us from direct fire but it made maneuvering difficult. Conlin was down and we couldn't get to him. He was about 20 meters to our front. McInvale was leading SGT Conlin when he got hit not five feet behind him.

McInvale was also hit in the opening volley, in the side under his left arm pit. Behind McInvale, close by was Darrell Kuhnau, a rifleman who was screaming and out of control. McInvale tried to reach Conlin but knew he was dead. He continued to fight, using his M79, firing rounds into the hedgerow to his front.

"1st and 3rd squads, move up and spread out and give me some covering fire," I barked out to the platoon. "Corum, get Price and Buckley to move the M60's up, one on the left and the other on the right," I yelled. "Hit the tree line low."

The VC returned a heavy base of fire themselves. Bullets were kicking up dirt and dust everywhere. You could hear them whistling by, sounding like mad bees. I could see muzzle flashes from three or four points to my front and set my sights on one of them. Wales and Tostado were dropping M79 HE's into the wood line. We got no orders from command or Partee about what to do. No effort was made to try a flanking movement to the left and into the tree line. Any attempt to move to the right with no cover would be suicidal. We were without an officer since 1LT Merrett was killed two days ago.

SFC Partee wanted some confirmation about what he was thinking. "Buckley, I want to call in a CS gas drop."

"Are you crazy? None of us have gas masks with us," Buckley said to Partee.

Partee then started talking about bringing in an air strike, but that thought got no traction. We were too close to the action and it would only endanger the guys trapped to our front. I was burning through my ammo pretty quick and others up front were also. I sent a runner to the rear to get us more ammo.

Our C.O. 1LT David Riggs seemed paralyzed and couldn't figure out what to do. He was our platoon leader prior to being elevated to the company C.O. back in September, but he looked like a deer in headlights right now. The Forward Observer called for a fire mission and within a minute there was a W. P. round hitting in front of us. The F.O. adjusted fire with a second round, then radioed, "Fire for effect!" and within a few minutes we started peppering the tree line with 105mm HE. That was all good for now, but we were still stuck out in the open and "danger close" to the artillery fire. The F.O. continued to adjust the artillery fire and was slowly walking it closer to our lines. Soon, we were being hit by the shrapnel. I asked my RTO, George Toto to contact the F.O. to adjust fire.

"6 X-ray, this is 35 X-ray over," Toto yelled into the headset over the noise of the battle.

"Roger 35 X-ray, this is 6 X-ray, over," the C.O.'s RTO shouted back.

"6 X-ray, have the Foxtrot Oscar (F.O.) adjust fire to 'up 50'; we are taking shrapnel, we are too close to the tree line, over."

"Roger 35 X-ray, will relay your request; what's your Sit Rep (Situation report) over?"

"Copy that 6 X-ray, we are still taking AW fire from three points. We have three whiskey papas (wounded personnel) down and can't reach them. Over," Toto stresses.

"Roger 35 X-ray, 6 X-ray out."

For whatever reason, the artillery fire did not get adjusted and the rounds continued to pour in. Right now, I was afraid that someone in the platoon was going to get hit. Again I asked Toto to get the C.O.

on the horn and to cease fire. He turned to me and said his radio is out of action and he's been nicked. His handset cord had been severed by small arms fire. The artillery firing continued so I jump up and raced back to where Riggs was. He was trying to light a cigarette but couldn't get the match to align with the end of his cigarette. I looked at the F.O. and told him what was happening up front. He finally issued an order to lift the fire. I raced back to my position, turning my ankle in the process as I sprinted across the plowed rice paddy.

The action continued on for a few more minutes and then the incoming fire ceased. Kuhnau was still yelling up in front of us. The platoon stopped shooting as well and we sat and waited and caught our breath. The sweat was pouring off us and we were all thirsty for a drink of water. When it looked like the enemy had retreated, we rushed in to get McInvale, Kuhnau, and Conlin. Big Jim had been hit pretty bad and needed help fast and a Dustoff. Tostado hoisted Big Jim up and we packed him out of there. Kuhnau was not hurt, but was really shaken up. There was no urgency for Conlin; he was gone. He took a burst across the chest and the front of his fatigue shirt was saturated with blood. 36 X-ray made a call for a Dustoff and when it arrived, Jim McInvale, George Toto, and Conlin's body were loaded on the chopper. We searched the area for bodies and found four VC KIA'd. Not long after that, we were picked up by Hueys and returned to our night laager at Pershing. Later, McInvale was cited for the Bronze Star for Valor along with a Purple Heart for this action.

George Toto, later, recalled that he played Little League against Conlin in his youth. All the way on the other side of the world and we found it is still a small place.

December 15th, B Co was ordered into Cu Chi for a 72 hour stand down. Elements from the 2/14th filled in behind us at FSB Pershing and conducted operations from there for the next several days. William "*Fox*" Hill took over 2nd squad, replacing Conlin.

During the months we were located on Hwy 1, with the exception of the operations in which we night laarger'd, there were very few ambush patrols. I don't have an explanation for this change in tactics. Maybe in part it was the weather and we were bordered by vast areas of rice fields which were flooded with water. But, that didn't stop our efforts when we were west of Saigon earlier in the year, so I have no answer. The reason I bring this up is the Starlight scope. This handy device allowed us to see in the dark. Today, everyone knows about night vision equipment, but in 1968, we safe guarded this instrument with our lives. It effectively transformed starlight into daylight.

Looking through the scope everything was cast in hues of greens. The more light that was amplified, the brighter green the object was. Using this scope was like using regular binoculars in the daytime. We never went out at night without one of these gadgets. Anyone who lost a scope had hell to pay and likely was given an Article 15, a dock in pay or paygrade.

The next day, the 16th, back at the FSB, my ankle was hurting and I could barely walk. I saw Doc Meyer who told me to take it easy so I kicked back for two days and stayed off the foot and limited my walking. I got a care package from Mom. When I opened it, there was a small artificial Christmas tree in there. I put it together and soon everyone wanted to have their picture taken by the tree. I didn't really feel in the spirit given the surrounds of where I was, but it was a nice thought.

We spent the next several weeks conducting road patrols out of Pershing along road TL-6A and providing security for the mine sweepers. On one of these days, we had a mission to leave the wire around noon. I, along with Partee, had briefed the NCO's, Keatts, Hill, Zimmerman and Corum on what our patrol would be doing. Soon, I put out the word to *Saddle Up* and start moving toward the exit in the wire.

Carlos Serrano, who had been rather vocal lately, blurted out, "I'm not going. I don't want to die. I can't take any more of this."

"Serrano, gear up and let's go. Nothing is going to happen to you." I told him. My words seemed to ring hollow with him. He didn't move off the bunker he was resting on.

"I'm not going, and you can't make me," was Serrano's answer to me, drawing a crowd from his squad.

"Are you refusing a direct order from me, Carlos?" I asked him.

"Yes," he said without looking at me as he fidgeted with his fatigue shirt.

"Last chance, Serrano, gear up or I'll report you to Partee for insubordination," I responded.

"I'm not going, so go ahead and do what you got to do," Carlos said in a low tone.

"OK, this will mean an Article 15 for you and you'll get busted in rank as well," as I moved away from him.

I informed Partee of the situation and he said we would deal with the issue when we got back from our mission. We left Serrano behind in the FSB. He would pay for his choice later. I believed he was reacting to the number of casualties we had so far this month. It had been hard on everyone.

No sooner had we been back at Pershing than we were sent back to Stuart at Trang Bang. I think this was because of the number of operations being conducted where we were loaning out companies to other units to shore up their tactical strengths, and this had happened to us as well. Occasionally, we mustered up a platoon size force and ran a few security patrols around the outer area of the FSB and kept the activity close at hand, just to make sure nothing was going on right under our noses.

Christmas was approaching and would be here next week. I checked the calendar to see what date it was, December 19th. It was night, around 1900 hrs, and third platoon was on bridge security. We were laying back and relaxing, kind of shooting the breeze as we often did late into the evening. Up the road at the FSB, we had our 8 inch gun support and some 105mm's. Sharing the encampment with us was a squadron of M48 tanks from A Troop as I mentioned earlier. Tonight the Cavalry was going out on night patrol, something I didn't think they would do. I am no *tanker* so I am ignorant of the tactics they employed. So, earlier, around 1800 they had rolled out of the FSB and headed south down Hwy 1 toward Cu Chu. Later that evening, around 2000 hrs, we began to receive RPG and 75mm recoilless fire from the tree line, along with SA and AW fire. The tree line was 300-400 hundred meters to our north across an open expanse of marsh and rice paddies and across the Trang Bang River which lies between the wood line and FSB Stuart. We could see the muzzle flashes and explosions as the RPG's and 75mm's were directed at the FSB and not the two bunkers that bracket Hwy 1 on the bridge which is east of the FSB by 200 meters or so and marked our position.

My platoon reacted to the fire and we grabbed our weapons and checked our loads. We had a .50 cal on the south bunker along with a 90mm recoilless rifle. Across the road we had a pair of M60's and the rest of our small arms. My bunker was the one on the south side of the road. Buckley, Price, and some of the weapons squad mans the north bunker. We don't really see any targets at this point. It was past sundown and pretty dark. The call was made for illumination and the FSB lit up the sky with aerial flares to the north and east. They made an eerie sound as they popped open and whistled from the starburst shell, then ignited hanging from a small parachute. Each flare would burn for five minutes or so. The artillery boys began to fire aerial bursts into the tree line. The 8 inch guns couldn't effectively do much because their rounds will have to travel about 1,000 meters before they can detonate, so it's up to the mortar crews and the 105mm crews to lay down some suppressing fire to discourage the VC from making a ground assault. We heard the crackling from their AK-47's, but they were still attacking the FSB. We fired off some rounds with our .50 cal, *tat-tat-tat…tat-tat-tat-tat*, keeping the firing slow but steady to keep from heating up the barrel, in case we really needed to defend our position from a ground attack.

The guys in the bunkers were growing restless because we felt like an island separated from the main force up the road. Yet, we were not in peril at this point. There was a lot racing through my mind. Were we finally going to taste a full scale night assault? The artillery was doing a bang up job on the tree line. The firing continued but we still didn't see any troops emerging from the forest. It was not much later and Puff the Magic Dragon arrived on station overhead. This was the C-130 equipped with three miniguns that can bring some serious firepower to bear. Puff began to lambaste the area where the enemy fire was originating from. We had been trading punches with the VC for about 30-40 minutes or so. We were yet to be tested, and maybe the VC were just probing our defenses for an assault later?

To the south, we began to pick up the Zeon lights from A Troop. They were racing back up the highway, and I could hear that old trumpet call in my head from many of John Wayne's movies as the cavalry arrived to rescue whoever was in trouble. We could now hear the rumble of the engines and tracks on the hard surface of the road. Every other tank had their cannon swung to the left or right to cover both sides of the road. Those tanks with their 90mm cannon swung to the right had their Zeon lights on and had lit up the countryside with a bright blue white beam of light. The light was so strong that you couldn't pinpoint where it was coming from if you were looking at it. They pulled up to a stop along the road between our bunkers and the FSB and began to fire back at the tree line. This added firepower along with Puff quickly broke the VC's ideas about continuing with the action and they quickly broke off the contact and disappeared into the night.

We had a great front row seat to the action. But it looked like we wouldn't be tested this day, so we relaxed and set down our weapons and lit up a smoke. We continued to gaze toward the trees, but all was quiet and it remained that way the rest of the night.

Whether it was Christmas Eve or Christmas day, our mission did not change. On Christmas Eve morning, A Co was in early morning contact up near the Hobo Woods. B Co was about two klicks north of Pershing, awaiting pick-up by chopper to reinforce A Co.

D Co, 2nd platoon, led by 1LT Toy Smith had road security. They had moved out from FSB Pershing to Six Alpha (TL-6A) and had started their security coverage of the 65th Engineers who were mine sweeping the road, moving slowly toward Trang Bang. As they were moving over the *Little Bridge*, a geological checkpoint on our maps, they were hit from behind by a machine gun. The enemy had let them bypass them and had set up at the bend of the road above the bridge. Part of Delta Co was over and past the bridge and then were hit with enemy fire straight into them from the right of the road. Delta machine gunner Bill Reid recalled, "Horton took a round through the arm…a rifle grenade had hit Dale [Potter] in the back of his leg, bounced off and exploded behind him; black and blue with lots of bruises and shrapnel."

Reid moved to Horton and patched him up with his *field dressing* kit. "In all the commotion I forgot how to tie a shoestring knot," Reid stated. "Here comes Walter *Twink* Ferguson like he was sliding into second base—sliding to a stop, tied the knot and was off." Reid rolled Horton into the side of the road and to relative safety. "I crawled back to where Mike and the M60 lay silent. Mike looked like he had taken a hit in the neck or upper shoulder. I turned my attention to the M60..." Reid fired his M60 machine gun until the barrel was cherry-red hot. He changed to a new barrel with his hand—badly burning his hand. Reid spent Christmas 1968 in the hospital.

CHRISTMAS EVE FIREWORKS...C Co was ambushed near the school house on Six Alpha and avoided any injury. Back at Pershing, according to SGT Larry Fontana—the main instigator—the 2nd platoon of Bravo Co broke into the battalion armament connex building, and stole a *BUNCH* of flares which they set off in a spectacular pyrotechnic display.

On December 30th, D Co was sent into Cu Chi for their 72 hour stand down.

As another footnote to my company's travels, the operation's reports reflect that the battalion continued to float back and forth between FSB Stuart and Pershing during the months of November and December. Each company every few weeks would find itself back at Stuart guarding either the artillery or the main bridge on Highway 1 for a few days or so.

On our next rotation from Pershing to Stuart, events did not go well. The NCO's were assigned the watches if the squad pulling duty did not. One night, after the platoon had run out of grenades, we were told to use plastic explosives. So, we started to use C4, a plastic explosive somewhat like playdough that could be shaped simply by squeezing and molding with your hands. C4 needs a combination of heat and shock to ignite. This meant we needed to use a fuse and a blasting cap. We cut a short piece of fuse several inches long and crimped a blasting cap to it then made a hole in a small block of C4 and inserted the cap and molded the C4 around the cap. The dimensions were about 1" by 2". The guard on duty would have the responsibility of lighting the fuse and tossing one of these made up bombs into the water every few hours.

Tonight around 2130, McKenzie, from one of the other squads, lit the fuse and in his haste to toss the lit C4, dropped it on the ground. McKenzie tried to retrieve the C4 and before he could do that, it went off, *KABOOOOM*. This loud explosion was followed by screaming and yelling. Everyone jumped to their feet and rushed over to the bridge to see what had happened. The site was not pretty, no pun intended. The soldier had lost his left leg at the knee, and the right one was damaged. He also lost his hand in the explosion. Blood was pumping everywhere and the G.I. was flopping around on the ground. We tried to pin him down as tourniquets were quickly applied to stop the bleeding and Doc Solomon did what he could to address shock from setting in. I jumped on the radio and called for a Dustoff which seemed to take forever to get to us from Cu Chi. I used a strobe light which I placed in my helmet so only the chopper pilot could see the flash from the air then talked the pilot down and we landed the chopper on the road, twenty-five meters from the bridge on the west side. After that accident, we stopped using any explosives on the bridge. I don't remember what happened to McKenzie. No one ever heard from him after he was dusted off.

With the arrival of December's replacements for the guys we lost, I now had a full squad at my disposal. Long gone was Sovey who was moved to another squad. I just couldn't use or risk using him in the field. Everyone had to keep an eye on him and not on doing their jobs. He always seemed to be in some sort of disarray. Given the choice, I would always leave him back at basecamp and go short handed rather than deal with him.

So my short staffed squad, still under the temporary acting assignment of Zoom, was made up of George Toto, Russell "Zoom" Zimmerman, Ed "Wally" Wales and Jesse "Taco" Tostado who was one of my grenadiers, Wales is the other. I finally get new replacements in Curtis Tassen from West Virginia, Joe Starling from Florida,

Clarence Carson from Alabama, and Joseph Collins, but can't recall where he hails from.

Zimmerman ended up making sergeant just before I left and I don't remember it was my recommendation to promote him or not. He's a good soldier who does what he's told and can be counted on to do his job in the field, but he turned into a real pothead and got high every night. The new guys of whom I don't remember a whole lot about were quick learners. My recall is faint because I only had them in my command for two months before I left the field.

We made 34 combat air assaults between November 1st and December 31st. Between these eagle flights, we worked the ground with the 2/34th Armor, 1/5th Bobcats, the 2/27th Wolfhounds and the 3/4th Calvary. Most of our movements were within a circular area around Trang Bang that extended out about 8-10 miles. The heaviest action was to our north and east. In October, we made contact with the VC 13 out of 20 days. There were many small skirmishes where we briefly exchanged gunfire only to have the enemy disappear and vanish quickly. They were not significant enough to write about but nevertheless, produced casualties, mostly in the way of wounded personnel.

1LT David Riggs who took over for McDaniel, when he went down on December 10th was replaced by CPT Melvin "Lee" Morrow, who hailed from Alabama. The change in command occurred at the end of December.

December replacements: Billie Bowden, Wetumpka, Alabama, Daniel Brady, Tonawnada, New York, Clarence Carson, Alabama, Pat Clardy, Sutherlin, Oregon, Collins, Dash, Jerry Dyer, Texas, Ronald Jochum, Nebraska, "Doc" Bob Meyer, Newburgh, Indiana, John Mikita, Oxford, Connecticut, Bruce Solem, Duluth, Minnesota, Joe Starling, Florida, Curtis Tassen, Ona, West Virginia and others, a total of 23 new replacements.

20. JANUARY, THE YEAR OF THE ROOSTER AND TET AGAIN

It was January 3rd, 1969 and a runner came and found me and informed me that we had an operational briefing at 1700 hrs. I checked my watch, and it was 1645. I cleaned up what I was doing and headed over to see SFC Partee. We had given him a nickname of Pappy soon after he joined the platoon, even though he was only 38, but looked 50. He always had a corncob pipe in his mouth and a Bible in his hand. I have not talked about Partee or Kaplan much. They went about their jobs as professionals. These guys usually lived for moments like these. Wars do not always come along often. They had a certain amount of allure to them.

It is hard to describe the camaraderie you get from sharing a bunker or foxhole with someone. It lasts a lifetime, the memories and friendships that is. War is something I would never want to go back through unless I knew the outcome and who here on this planet has that kind of vision to see into the future? Half the time, you're scared shitless, yet confident, and ill at ease with each day. The rest of the time I spent trying to get through the day, one day at a time, and being thankful I would see another sunrise. There is an element of raw power and finality to everything undertaken. It sucks your emotions dry and elevates the heart rate. Your reflexes are razor sharp and your senses operate at optimum levels. War has an ugly beauty to it, as long as you are not one of its victims.

There was a call for the NCO's and I headed over to our command post and said hi to Keatts as we waited for Corum and Hill. At 1700 sharp they arrived and the NCO's were assembled. We were briefed on tomorrow's mission as we listened in on the operational outline.

We left the wire at 0800 hrs and conducted a RIF to the village of Rung Cay. This was a company sized force and 3rd platoon had the point. The weather was pleasant as it was most days this time of year. A person could wear one layer of clothing and be comfortable day or night. The sunlight was bright and warm as it beat down on us. The skies were clear of cloud cover and had that rich blue tone not normally seen this early. The countryside was quiet and still. There was very little activity. Not much pedestrian traffic that we could see either.

The peasants usually headed to the market place to trade their goods or animals for staples when they were not tending to the fields. It was amazing to see how much they could carry as they rested a pole that had two fully loaded baskets suspended from each end, hoisted upon their shoulder. The sounds of motors running from their Vespa trucks and vans competed with the whine of the Honda motorcycles as they spewed a cloud of oily blue exhaust fumes into the air. The Vespa would be totally loaded down with passengers in the back, like an open framed bus, with seats and all. The top of the Vespa had cargo space. Lashed to the roof would be pigs, chickens, ducks, and other goods, all ready for the market. As these vehicles worked their way down rutted dirt roads, they would sway to and foe, almost tipping over from being top heavy.

The civilians on motorcycles could manage to balance and carry their wife and fit several children on somehow. It was not uncommon to see a rider with a passenger on the back holding plants, doors, cages of animals etc, and nothing was tied down, just performing a balancing act for all to see.

Most peasants were living off the land any way they could. They raised rice to trade if they had the land for it. Otherwise, they grew vegetable gardens and had an occasional fruit tree around. It seemed most families owned at least one water buffalo. Those animals were big and didn't have a good feeling about Americans. The kids were in total control with these beasts. They moved them around for their papa-sans and moma-sans. The buffalo were usually kept in a stall or small corral near the family's grass hooch. Along with the buffalo or several buffalo if the family was wealthy, they had an assortment of pigs which were used for food or trade. Chickens ran free in the surrounding plants, trees, and grasslands. The makeup of these family units included the grandparents, parents, kids, and other relatives who squeezed into the available floor space.

In the area that we were sweeping through, the houses were scattered and separated by hedgerows of trees and bushes. We cautiously skirted the open areas and always kept some cover near to our line of travel if we could. Exposing ourselves out in the open was asking for sniper fire. There were small ox cart trails that connected the villages and foot paths that took the farmers back and forth to their fields. We avoided moving directly on any traveled surface. Maybe it was safe but we did not want to be predictable in our movements. The morning was slowly passing by and we completed our sweep through the first few checkpoints. The RTO reported our progress to the battalion. They needed to know our exact location at all times. We circled around behind another village and entered it with the sun to our backs. Everything looked normal to us. The local residents didn't seem to be on edge and the kids were hanging out around us trying to get us to buy a soda and ice for fifty cents. If we bought ice, the kids would have to peddle off to somewhere where they would get a small block along with the soda and return. The fastest way to cool a can of soda, since it was the same temperature as the outside air, was to lay it on the block of ice and then spin it. It would carve a trench into the ice and in no time, was cold as the ice block. Yea, they were drug dealers too. Some of them were selling pot. The company wound its way through the village and we left as quietly as we arrived.

January 5th, Charlie Co conducted an air mobile assault with no results. On January 7th, we air assaulted into XT535224 and from there we made two more assaults before we set up a NL site at XT547257. It was a quite night. The sun had set and we began the process of watching and waiting to see if the enemy would attempt any kind of action against us. PFC Jerry Dyer, a Texan who was in 1st Plt and had arrived mid-December, recalls that night.

"The company was scattered around the perimeter by platoon as usual. I was set up along with my partner in a fighting position. We were doing our best to create some kind of cover. To my right were two members of the squad who happened to have night watch and had the Starlight scope. For the first hour or so, all was quiet. Then the conversation started up."

"Holy shit, I see two gooks out there!" the guy with the scope said to his buddy.

He peered through the lens of the scope again. Lowering the scope and squinting his eyes, he looked into the darkness. Nothing! There was nothing there. Again, he raised the scope and looked again. Yep, there were two gooks and they were very close.

"Shit," he said, but again when he lowered the scope, he couldn't see anything.

His partner asks him what was he looking at and grabbed the Starlight, but he too had the same problem. One second they were there and the next, gone. Dyer, hearing the excitement in their voices, maneuvered his way over to their position.

"What's going on?" Dyer asked, "What's all the fuss about?"

"We've got two gooks to our front in the scope, but I can't see them and they are really close!" was the answer.

Dyer said, "Let me have a look see," in his Texan drawl, then raised the scope up and peered out.

"Damn, you're right, but there's three of 'em now!" Dyer lowered the scope and looked hard into the night with the other two grunts, but, no gooks. They couldn't see anything.

"What the f#@$ is going on?" they asked each other, knowing that the enemy seemed to be close enough to reach out and touch. With all of this commotion going on, 1LT Louis Gietka showed up. He was leaning over and behind the group.

Dyer once again raised the scope but now he didn't see three gooks, there were four of 'em.

"Shit, mother fu#$ker, they're right on top of us!" he said.

Suddenly it dawned on Dyer what was happening when the Lieutenant knelt down and he started laughing. He was beside himself. The moon had come up after dark and everyone's silhouette was being projected to their front from behind them and they were seeing themselves reflected off the hedgerow through the *Starlight* scope.

"We all had a good laugh over that one," Dyer recalled.

The next morning, we made one more assault landing with negative contact before being lifted back to the FSB on the evening of the 8th. A Co was sent into Cu Chi for their stand down. I had a chance to go see Bob Hope and his Christmas show, but decided to send several guys from my old squad. George Toto was one of the guys picked to go.

Several days earlier, I learned that I was to go before a promotion board for the possibility of making Staff Sergeant. It was January 6th and today was my appointment. The board was manned by four officers, the company commanders and one from battalion. They asked a number of situational questions and wanted to know how would I respond to each one. We were sitting outside in the sun during this process. The sun was beaming down on us and visible beads of sweat were appearing on the foreheads of the board members. The Q & A lasted about 30 minutes, more or less. They would then assess my answers and decide if I was promotion material or not. I thought everything went well, so time would tell the story. The interview was also about me being asked about being a platoon sergeant. I was already acting in that capacity, but for how long? Those questions were easy to answer, given what I had already accomplished.

A number of days passed by and our replacement platoon leader arrived. I believe it was around the beginning of the second week of January, but my memory might be playing tricks on me. He was 1LT Michael Sheehan from Moosup, Connecticut. He stood about 5 ft 8", medium build, had dark brown hair which swept down and across his forehead. He had a mustache which complimented his features. He was fresh out of OCS (Officer Candidate School) at the age of 19. He spent one year at Ft Jackson, six months as an instructor, and six months as a AIT training officer before receiving his orders for Vietnam arriving as a young 20-year old, and showing his eagerness to get into the action. He had been in country since November, arriving around Thanksgiving, where he reported into Cu Chi for orientation and recalled having served turkey dinners there. He was assigned to other duties before being sent to FSB Stuart, and being assigned to 3rd plt as the platoon leader. I, along with Hill and Keatts took turns sharing our knowledge about combat operations with him and how this game was played and what was needed to be done to stay alive. We reviewed our normal routines and duties with him while he soaked in the conversation. It was our intent to create a good line of dialogue in the process so he would know who we were as well. Sheehan, with his arrival, was the third PL for us in the last two and a half months.

The word came down from Brigade that the 2/12th had the mission of destroying local forces of the 109th and 268th VC/NVA Regiments and local VC cadre in the upper Citadel area, which was at the top of the Hobo Woods and which continued to be part of our normal area of operation. But then, we were told a lot of things and they don't always come true. Just more army scuttlebutt, or was it?

It was now January 11th, and we were on road security once again. By this time, we had almost made trails along the sides of road 6 Alpha. Moving away from the perimeter of FSB Pershing, I yelled out to "*Lock and load!*" and everyone chambered a round and checked the safety position on their M16's. Today, Doc Leavy Solomon, a black kid from Georgia joined our merry band of brothers for a stroll in the country. He had been with us since September and was assigned to our patrols often. He had a big smile and wore black rimmed glasses. He was funny and had lots of stories about his growing up back home. The first part of the patrol was fairly easy as the terrain was open and uncluttered. But out a little further, we began to maneuver our way through the rubble of once stately brick and mortar houses. These homes must have been a sight to see when they were first built. Now you just can't visualize how this country must have looked before all the wars and destruction. My squad, along with SGT Keatts' 1st squad, was moving along the east side of the road toward Trang Bang near the old school house. SGT Hill's 2nd and SGT Corum's 4th squads along with LT Sheehan made up the rest of the patrol along with Doc Solomon, was across the road on the west side, moving in parallel. We were out ahead of the 65th Engineer's that were sweeping the road for mines. Along the way, Doc Solomon was doing some MEDCAP work with the local villagers as he walked with the platoon.

My squad had the point and we were several hundred meters ahead of the sweep team. It was slow going walking then waiting for the engineers to complete their mine sweeping along each section of the red dirt road. We paused to wait when there was a loud explosion to our rear. We saw a column of black and red smoke from the blast, quickly rise up into the sky.

"What just happened?" I asked as we held our position. Richard Coleson, my RTO was listening on the radio.

"Doc Solomon stepped on a mine. He's dead," says Coleson.

1LT Sheehan recalls, "The guy that I knew as Solomon was in the road on a MEDCAP, serving the Vietnamese who came to the road as the engineer's support vehicles followed behind the mine sweep team. After he stepped on a mine, I went to the road with my RTO and a few others. There was a large crater in the red earth, from a sandbag mine intended for a vehicle and probably similar to the one that got me several months later. Solomon's corpse was simply a red clay-covered upper torso with a head and nothing else—no blood—the features of his face disguised beneath the soil—just a clay plastered figurine. It was a hauntingly surrealistic image that remains crystal clear in my mind. Those sandbag mines were often triggered by a homemade little bamboo box made from a couple opposing segments of bamboo wrapped with bare wire. I don't think that there was a lot of quality control. The devise required a large bag of ammonium nitrate fertilizer, a little C4 plastered around and electrical blasting cap buried in the middle of the fertilizer, and a d-sized battery or a couple cells of an AN-PRC-25 batteries. It seemed to me that there was a disabled Jeep on the south side of the crater, but I can't be sure of that, as my mind apparently focused on the image of Solomon."

Just like that, death strikes out. It was quick and merciless this time, but painful none the less. Why did the medics have it so tough? He was a good kid who did his job without fear and reservation. Like all the good Joes, he will be missed by us. We radioed for a Dustoff and waited to have Solomon's body picked up. After that, we resumed the patrol, this time even more vigilant about watching where we step.

The days seemed to be passing quickly as my attention began to turn toward ending my time here and returning home. I had two months left on my Southeast Asian tour. Partee had fallen ill and with that, I found myself still running the platoon as the senior NCO. The date was January 12th and we were still shocked about Doc's death.

We had a combat assault schedule for the morning at 0740 hrs. Battalion had a short flight north to the village of Sa Nho (2) for Charlie Co. From there, we patrolled south two clicks and ended up at a PZ where at 1300 hrs we were then flown south and landed west of the village of Rung Cay. We searched the area and in the process someone from another platoon set off a booby trap which resulted in us calling for a Dustoff. I don't know how serious the wounds were or who was injured. After that diversion, we returned to Pershing for the night.

Charlie Co was given a day of rest to sort things out and take care of the usual activities one does when time was on our hands. We got cleaned up by taking a shower, shaved, wrote letters home and listened to the radio. Somewhere in that long journey of repeated events that seemed to be endless in nature.

Billy Graham stopped in at FSB Pershing for a visit along with a bunch of polished boots, meaning rear echelon officers who paraded around in their starched fatigues and clean boots, hoping to impress someone. It wasn't us, believe me. We went about our business while they attended to theirs. Some of the guys went over to hob nob with Graham, as for me I just shrugged it off.

LT Sheehan was learning more with each patrol we went out on, but at times, had a stubborn streak in him of which he would admit to later. One morning while out along TL-6A, 1LT Sheehan thought that he had discovered a .51 cal machinegun site. Buckley told him, in his opinion that it was nothing more than a slit trench, a latrine. Sheehan wanted to blow it up, still believing he had discovered an enemy firing position. Buckley lost the argument and Sheehan tossed in a grenade, only to discover Buckley was right all along. He didn't live that one down for a long time.

January 13th, we were flown southwest from FSB Pershing, and directly below Go Dau Ha just east of the Vam Co Dong River, nicknamed by some, Oriental River. The company conducted sweeps of the local villages, Giang and Thanh, looking for caches of any sort. This was low country, mostly marsh type land with lots of water, reeds and leaches, and rice fields. After conducting a search of the area, we were flown back to FSB Stuart for the night. We returned to the same area the next morning when we were picked up at 0925 hrs. After conducting another fruitless search at 1555 hrs we were picked up and returned to Stuart again for the night.

There must have been something going on that intelligence had picked up on because we did not operate in this area southwest of Trang Bang. It was really out of our area of operation and control. The next day, after a briefing from Sheehan, we were told another combat assault was scheduled again for the same area for the 15th. This time, Charlie Co was going to be joined by the ARVN's who would fly in behind us and we were to conduct a sweep east along the Trang Bang River. The ARVN's didn't really have a stomach for fighting the VC unless they were cornered. It looked like a game to us, that they played. The day was spent on patrol and the intelligence we had, came up blank.

At the end of the day, we returned to FSB Pershing. On January 18th, CPT Richard Wiggins was moved to battalion and took over as our S-2 officer. Replacing him, CPT Layton Matt, who was there a week, then CPT Charles Boyd from Shreveport, Louisiana, took over for the next four months. He would turn out to be a good C.O. following in Wiggins footsteps.

The supply Chinook had brought in a fresh water trailer, more food supplies, and fresh fatigues for the company. That was always a treat, getting the chance to bathe the stink and dirt off our bodies. We didn't have shower stalls or any elaborate means of washing. I grabbed my steel pot helmet and filled it with water and proceeded to give myself a sponge bath minus the sponge. It felt really good and pulling on a fresh set of fatigues was even better. I turned in my dirty pants and shirt. They were so saturated with sweat they didn't want to dry out anymore. I brewed up a cup of coffee and grabbed a smoke and went looking for Price and Buckley. They were at the other end of the platoon area chatting with Hill, "Satch" Satterswaithe, Gooding, and Clardy. The radio was on, well, it was always on somewhere and was playing *Crimson and Clover* by Tommy James and the Shondells. We started clowning around and out came the camera and we started taking pictures of the group. After a while, we got a little goofy with our poses. It was all in fun and kept our spirits up. Some of the guys had cameras, but generally they were not taken out on operations. Today marked the last day for me as platoon sergeant. Sheehan was in command and SFC Partee returned to his duties as platoon sergeant after being ill for some time.

On the morning of January 16th, the company conducted a RIF north of the FSB to an area west of the village of Cau Xe. We engaged in a minor skirmish with several VC who eluded us for the time being. But later in the day, we jumped them again, only this time with the aid of several gunships, we took them out. Somebody also managed to trip another booby trapped grenade which resulted in another Dustoff. Every week, it was the same thing repeated. The activity levels picked up and we continued to fly out of Pershing on assault missions around the area. This whole process was an attempt to catch the VC out in the open. On the 19th it was another mission with the ARVN's, again to the south of Trang Bang but no luck. The company took a few days off getting rested up on a scheduled stand down in Cu Chi. It was another chance to get our minds off the last few weeks of combat. Some of us, Zimmerman, Fowler and I headed over to the NCO Club and got a few beers.

The Vietnamese had their own beer that they brewed called Ba Moui Ba which is Vietnamese for Thirty Three or '33' beer. This Da Nang-brewed golden lager was produced by France's *Brasseries et Glacieres Internationales*, until the plant was nationalized after the fall of South Vietnam in 1975. After nationalization, Vietnamese-made beer was excluded from most major export markets other than Japan for years, and the French continued to produce 33 outside Vietnam under worldwide license. The Vietnamese beer became known as 333 (or *ba ba ba*), and the 33 vs. 333 dispute plagued Vietnamese brewing for decades. Heineken's Saigon Brewery Co. now produces "33". Lots of beer drinkers tried this beer, but I stuck to U.S. stuff.

After our stand down was over, we patrolled TL-6A, and then back to the choppers again for a few more flights but zero results. January 24th was just another beautiful day in the countryside and we were sweeping out to the east of the FSB. Not too much to report on this day. The lead platoon, the 1st jumped a VC who attempted to run from us and was gunned down. Other than that, it was a quiet day and no further action took place. This area that we were patrolling was always a hotbed for us and engaging the enemy was always expected.

That evening, LT Sheehan sent for me. I walked over to his bunker and reported in. I was told that HQ wanted to know where I would like to be stationed when I returned home and where I would like to end my service commitment. He said to write down three places on my *Dream Wish List* and return it to him in the morning.

That was cool, I'm thinking, but where? I wrote down Fort Ord, in Monterey, California, the Presidio in San Francisco and last I put down Fort Lewis, Washington.

On the morning of January 26th, CPT Morrow's company engaged a squad of VC operating out of a tunnel complex near the village of Rung Cay. They tried to take us on with small arms fire. We had them outgunned and killed three of them quickly. Within a few minutes, four VC surrendered. One of them was wounded and we called in for a Dustoff chopper and after doing a search of the prisoners for weapons, we tied them up, blindfolded them, and got them on a chopper headed back to Cu Chi and interrogation. We had two wounded in the engagement and they were also lifted out.

From January 27th to February 5th, Charlie Co took turns handling road security detail along 6A, escorting the mine sweeping teams and keeping the VC away from the supply route that led into FSB Pershing. No one was seriously hurt during this period.

During the morning of January 28th, the company was flown out to the east for a combat assault south of the village of Sa Nho, a nasty area. But the day proved to be friendly toward us. We made our sweeps and found nothing out there and returned to Pershing.

Our new replacement, PFC Bruce Solem, was cutting his teeth by walking flank as we were doing road security on 6A. This is a quote from him, but he is mistaken about where we were. At the time, we were conducting operations out of Pershing. Bruce Solem remembers back to that day. "It was January 29th and my first day walking flank. We took a short break and it was quiet until I got knocked ass over tea kettle by that RPG. Some of the guys were telling me that I had a million dollar wound but I was back in a few days. We were working out of Stuart that day. I'm pretty sure we were just off the blacktop in that first piece of woods east of Stuart on the north side of the road." Solem took some shrapnel to his hand, but would be OK in a few days. He was kept out of the field until his wound healed. I was starting to watch the calendar more.

LT Sheehan wrote, "Several days later, we were ambushed again near that same spot. The VC used to pick up the U.S. units when we left Stuart. They'd run along several well-worn trails that ran east of Trang Bang to intercept and ambush us after we left the village. On the day of that particular ambush, I had pre-planned mortar fires along the hedgerow that was intersected by those trails; but, at the last minute the order to fire was disallowed by the Vietnamese Liaison Officer at our Tactical Operations Center. Apparently, they presumed that we were pinned down. While I ranted and fumed, the Vietnamese dispatched troops from that little compound that was situated along the highway between Trang Bang and FSB Pershing. A lieutenant of the Vietnamese Regional/Popular Force allies was killed during that assault to save us."

The morning of the 31st, my squad was assigned road security detail and we were to escort the 65th Engineers as they did a mine sweep of the road. They had a tough job and the road security thing made us a prime target for snipers and booby traps. On our way to Trang Bang, we were passed by an M48 tank. He didn't get but a 100 meters past us when he detonated a 40 lb mine. That caused some extensive damage to the tank, blowing the road wheels off the undercarriage. I watched as a set of wheels flew about 100 feet into the air. The tankers were OK, but shook up. One crewmember was bleeding from the eardrums. We put out a security perimeter and waited until support arrived to deal with the damaged tank so we could carry out our mission.

Many of us remember our time spent in and around the small town of Trang Bang. The unit spent close to 17 months in this area. I did not know it at the time, but this would be the last time I had ventured through its streets. I use the term 'town' because it had defined streets, although they were still dirt until the Army Engineers arrived and laid down gravel and eventually paved some of the major roads. Whether this was done to make them more all-weather roads or some attempt to minimize the VC's efforts to place mines in the paths of our convoy's and road security patrols, only they have that answer.

Hamlets and villages were the smaller cultural centers for the rural farmers to gather to exchange some commercial trade. In Trang Bang, we saw a city with mortar and brick buildings, some corrugated steel roofs, and a few well-placed signs advertising their places of business. There were plenty of street vendors both inside the city limits and stretching slightly beyond, selling everything from flowers and hot food to soda and ice.

Most of us would pronounce Trang Bang as TRAANG Bang, but the Vietnamese pronunciation is Tron Bon, because the "g" is silent. Located within the town gates to the east side was the ARVN compound, a small circular shaped affair with its rows of concertina wire and guard tower. How many times did I recall seeing ARVN's in uniform, packing their weapons and wearing flip flops? They looked ill prepared for the dangers that lurked around the corner. Half the time, they didn't even have a magazine in the weapon.

Trang Bang was the intersection hub for several main roads. From the south was Hwy 1 which connected Cu Chi to the city. To the north was the gateway to Tay Ninh and Dau Tieng and to the west was the road to the Cambodian Border. It was here also that we fondly remember 6 Alpha. The road began in the center of town at the intersection of two other roads. One road reached Tay Ninh and the second road ended in Dau Tieng, also called Tri Tam on the maps. Leaving Trang Bang, 6 Alpha, a hard packed red dirt road in the dry season, took a northerly route reaching the Hobo and Boi Loi Woods and beyond. 6 Alpha turned out to be both a blessing and a curse.

Where TL-6A split off from the main road was a downtown

market area of shops and restaurants. It was a bustling location full of villagers, old and young and children. Vehicles had to slow down in order to navigate through the heavy pedestrian cross traffic. The kids would line the street to wave and shout at the passing G.I.'s. "You numba one" or "you numba ten G.I." They would ask for cigarettes and offer to sell you sodas and beers.

I'm not sure when FSB Stuart II was created, but I do remember being there in May, 1968. It was located on the west side of town on Hwy 1 toward Go Dau Ha and the 2/22nd and 3/22nd were operating road security out of there. It was abandoned some time later and resurrected as FSB Stuart III on the east side of town, below the Buddhist temple and about 200 meters from the Hwy 1 bridge that spanned the Trang Bang River which flowed into the Song Vam Co Dong. In Vietnamese, song means river.

The most famous photograph perhaps of the war, was of Kim Phuc, who lived behind the temple at the edge of town. She was the little girl who had her clothes blown off her by the ARVN Air Force who were napalm bombing the NVA and VC who had taken over the city. This was well after we had left the country in 1972. The photo made the cover of LIFE magazine. I had the opportunity of meeting her granddaughter when I visited the area in November, 2010. She lived just behind the temple on the main street.

Today, FSB Stuart is the site of a power substation and the rickety old bridge on Hwy 1 is now a four lane concrete structure. The temple still stands at the edge of town and many of the rice fields still exist. It was rather strange standing on the new bridge, starring down at the brown colored water as it slowly drifted down stream. I could visualize the caged white geese we had placed slightly upstream as an early warning system.

At the beginning of October 1968, when FSB Pershing was established, we relied on our supplies being delivered by the 242nd

ASHC Muleskinners and their CH-47's. Later this operation was abandoned and road TL-6A or Six Alpha as we called it became our main supply route back to Hwy 1 and ultimately back to Cu Chi and the Division basecamp. The resupply convoys were made up of Jeeps with .50 cal MG's, usually at the front and rear of the line, deuce and a half's, a term describing large trucks with dual axles and canvas covered truck beds, with an occasional 3/4 ton truck. There were also mechanized units every so often, along with the local villagers, all sharing this road. At first, all was quiet, until the VC realized that we were going to be using that route almost daily and that is when they started mining the road. After that, road security duty, escorting the 65th Engineer Battalion each day to sweep the road and look for signs of booby traps was a task normally assigned to a platoon sized force.

Sniper fire, booby traps set along the edges of the road, and occasional ambushes were fairly common. We lost a few good men on 6 Alpha. Three men from my platoon were jumped at the edge of Trang Bang just beyond the ARVN compound in December, 1968. All were killed in a hail of gunfire and RPG's. Doc Leavy Solomon died when he stepped on a mine just past the bridge out of Pershing and later, SGT Dees was killed in the same area in an ambush near the bridge on 6 Alpha.

In talking to some of the guys who served in Echo Co which was formed September or October 1969, I was told that there was a compound located somewhere in Trang Bang that had barracks with cement floors. There was an underground concrete vault for communications. Officers and uniforms were scarce in the area. The compound housed both Provincial Reconnaissance Units and U.S. personnel. These men worked in RECON and were part of the Operation Phoenix group. I am not sure if everyone was involved in that counter intelligence group or not. You can read about Operation Phoenix on your own if you *Google* it or go to the website at www.212warriors.com. It was a rather unsavory job in which I was not sure exactly how far the U.S. was willing to push the limits of warfare, but apparently, this particular mission did test the line and crossed over. I'll let you be the judge.

Today, Trang Bang is a bustling place with motorcycles and cars going in all directions. The market place is smaller, and not the central hub it once was. It was chilling walking down 6 Alpha and reflecting back on all that had happened here so long ago.

January Replacements: Dale Hall, Minnesota, David Abenathy, Plevna, Kansas, Dale Studebaker, plus a few more for a total of eight to the company.

21. PATROL BASE GRANITE — AND HOME

February 1st, William Hill, Elmer Lightner, and I received orders awarding us the Bronze Star for meritorious service in connection with military operations against a hostile force. Only a few more days to go and Hill, Price, and I would be pulled off the line as we would reach 'Short Timer' status. A term we overheard in conversations and saw on some of the camouflage helmet covers was *short timer*. It was a good thing for the G.I. who was one. It meant he was counting down the days until his DEROS date or end of his tour of duty. DEROS means "date estimated return from overseas." A short timer had 30 days or less until he climbed on board the *Freedom Bird* (Boeing 707) and flew home to the *world*.

The battalion had an unofficial rule about taking the troops out of combat situations once they get within 30 days of their rotation back home. My DEROS date was set for March 10th. The guys were ribbing me and a few other guys who were also *short timers*. Once again there was a change of command ceremony and we got a new battalion commander and his name was LTC J.P. Grisham who replaced LTC Thomas Dreisenstok. Thomas held the post from October to January.

The morning of February 6th C Co remained in the FSB. We were notified that we would be going out at 1330 hrs and were to be airlifted to an area around XT586290 to conduct a RIF. Little did we know that we would set up a night laager position and spend the night out there. CPT Morrow had us move through the area cautiously, scanning the wood lines as we approached or skirted them. We expected the enemy to take his best shot at us at any time. There was always the uncertainty that the next time we got close to any type of cover that the VC would be waiting to shoot point blank at us.

For the veteran, there is the aid of that internal instinct that raises our level of awareness that tells us when something bad is about to happen, but not this mission and we did not find the VC. We set up a patrol base on the northern edge of the Hobo Woods, later referred to as PB Granite.

Just after midnight on February 7th, around 0040 hrs, back at FSB Pershing, they started getting some incoming rounds from the local VC. Those rounds were hitting our company's area in the base. The 1/8th Battery B artillery responded by returning fire along with the Four Deuce mortar crews. The basecamp got hit with 9 rounds of 82mm mortar, several rounds of 120mm mortar, and a 122mm rocket. The rocket hit the artillery area and we had 12 wounded, requiring a Dustoff.

Later that morning, at 0828 hrs, the remaining elements of Charlie Co, along with elements from Delta Co led by CPT Paul Allen who resumed command from CPT James Ellis, were airlifted out to the patrol base. Once we hooked up with the rest of Charlie Co, we conducted a sweep to our east and returned to the NL that afternoon. We were in some really dangerous real estate and expected that our time here wouldn't be pleasant for either the VC or for us. There was just too much opportunity for something to go wrong. On the 8th, Charlie Co was at the Hobo patrol base and flew three clicks northeast to a landing zone just south of old highway (TL-15)—halfway between the Thai Thai stream and the village of Xa Duoc on the Saigon River. Charlie Co moved northwest towards Duoc and turned left just short of the small temple in the town. We didn't make contact so we moved southwest to a PZ and returned to base.

That evening I got a hand written note from our NCOIC, Hiram Marziano informing me that I was going to be assigned to Fort Ord in Monterey, California, to close out my military duty. It was three hours from home, so I envisioned going home on weekends, or so I hoped. I was content to find out that I would be going to Fort Ord.

The morning of February 9th we were working with the CRIP and A Co from the 1/5th Bobcats. The CRIP's were in a blocking position after Bravo Co and Alpha Co from the 2/27th Wolfhounds had joined the operation. That afternoon Bravo made contact with an unknown number of VC and called in gunships and artillery for support. Bravo Co suffered three wounded but no casualties. Later that afternoon, Charlie Co working with Delta Co came under attack and we repeated the same routine by calling in supporting artillery fire and more Cobra air strikes. That engagement lasted several hours and the action was more towards Delta Co than Charlie. The result was four U.S. KIA's, two WIA's, and one VC POW. The company spent more than a week in the field working with the Bobcats conducting sweeps of the area. For some guys, they did not want to be around the mech boys because they were such a magnet for the VC, but for others, it was the muscle of the Army and the sense of its firepower got the testosterone flowing.

As for me, I was pulled out of the operation and along with Barry Price and William Hill, we were ordered to remain at the patrol base. Our time had come to an end for patrol duty. I had mixed feelings about this, wondering if my squad and the platoon could handle our absence, but the answer was, they would move on without us.

I was to coordinate our resupplies and control the landing area.

Since we were so close to our DEROS date, no one wanted to see us take a bullet and not make it back home. So, we did what we were told. We kept an eye on the perimeter, read anything we could get our hands on, and waited for the unit to show up at the end of the day. We listened to what everyone had to say, reliving the day's activities, offered our advice and counsel to our replacements on what they needed to be doing. The time went by slowly as we waited to report to Cu Chi and then to Long Binh to fly home. I wrote home to Mom and Dad telling them of my estimated return date and to inform them not to write anymore letters after the 25th or I wouldn't receive them.

A week and a half before we departed the field for good, we left the patrol base and flew back to FSB Pershing. We got one last and final salute from the VC when they decided to mortar FSB Pershing around midnight. When we moved into FSB Pershing, we had to construct our own bunkers as firing positions and also as places of protection. When I constructed my bunker, I placed three layers of sandbags on the roof over three pieces of PSP (perforated steel plate). It felt good knowing I had a nice roof over my head. Anyway, as these mortar rounds were hitting, they were getting rather close. One big one landed on top of me. I didn't know how close until the next morning when we had some daylight. It was a 120mm mortar round and it hit about three feet on the other side of my splash wall behind my bunker. I dug up the tail fin and ended up taking it, along with an 82mm tail fin from the same shelling, home with me. It was a nice souvenir and reminder of my close call.

As for my last days of combat from January 1st until February 9th, the company conducted 17 air assaults in the provinces of Hau Nghia, Tay Ninh, and Binh Duong. I do not have clear recollections of when or if I had a chance to say good bye to my fellow soldiers. It may have been just a timing issue where they were gone in the field and my time had come to leave the field and report in to get ready to depart this land of many faces. I shed no tears seeing this place disappear behind me, but I would never escape the memories created during my time in Vietnam.

LT Sheehan recalled: "SFC Partee headed back to Cu Chi around this time, mid-February, to heal some back injuries. He returned to the field a couple times after that, but didn't go on RIFs with us until his final return, in early March."

While I was waiting and killing time at Pershing, on February 28th, I received orders promoting me to Staff Sergeant (E6) effective February 13th along with Doc Dennis Sheppard. Russell Zimmerman made Sergeant on the same set of orders.

It remained quiet around the basecamp, but our company was finding plenty of action now that I was away from it. In the middle of all of this, LT Sheehan and Russ Zimmerman butted heads over the distribution of ice, and a few swings are taken. Sheehan was written up for an Article 15. Zoom (Zimmerman) was close to going over the edge by this time, having spent too much time with a bong pipe at night. I questioned his focus.

"I don't think that I got more than two weak punches in before I was grabbed, but I received a Battalion Article 15 for that fight with Zoom—Zimmerman might get a kick out of knowing that," Sheehan remembered.

"I've long since either lost or destroyed my copies of my military records, so I can't check the signature on the documents, but I think the Article 15 was signed by a Major Hutchinson, or Hutchins, or something like that. Before he sent me back out to Pershing, the kind Major also threatened to honor my foolishness by reassigning me to MACV with a MAT Team. I was blown up before I reached the FSB, so he didn't get a chance to make good on that threat—also, I don't think they actually took the two weeks' pay that I was supposed to forfeit. I did end up with MACV, but it was a couple years later when things weren't quite as hot; there were no MATs left, and I was part of a two-man advisory team in Duc Thanh District—one of the nicest cultural experiences that I ever had."

I am worried about the company, my platoon, and squad after hearing that they were in some real contact with the VC. The news was bad and was troubling to me during those last few days in country. The month of March was characterized by continued heavy activity in all Brigade areas. The newly executed offensive had gained full momentum. The Division intensified its search for the enemy through extensive ground reconnaissance, air-mobile operations, and widespread aerial reconnaissance. This approach brought success as elements of the Division began to hit the enemy in his staging areas.

Craig Schoonderwoerd, over in 1st Plt was about to DEROS back home. He was on a mission and was asking anyone if they wanted him to contact our parents when he got back to the States, just to let them know we were doing fine. I gave him my parents' address in Cloverdale, California. I don't recall ever doing this, but years later when Craig and I got together for reunions, he showed me his note of hand written addresses and instructions and there at the top of the list was my parents address. I really got a chuckle seeing this document once again 45 plus years later. Craig is a wonderful story teller, and his recall of events rekindles old memories.

LT Sheehan wrote, "SFC Partee returned from Cu Chi and when he rejoined us, he had to lead the platoon while I went back to Battalion Headquarters to receive my Article 15 for the "fight" with Zimmerman. As I awaited the convoy, the platoon ran into stiff resistance in the Bo Loi's, or maybe it was Sui Dat 1 or 2. I listened on the radio as Lynwood Keatts called for supporting fires; presumably Partee was injured at that point. Frankly, I bear some serious guilt over my role in his death."

March 4th, the company found trouble described in these two accountings of the day:

On 4 March 1969 elements of 2d Battalion 12th Infantry engaged an enemy force at grid XT551225. A/3/13 Artillery and B/1/8 Artillery supported the 2/12th Infantry with 533 rounds. B/1/8 Artillery was credited with three secondary explosions. At 1300 hrs, after other elements of the 2/12th Infantry and 1/5th Mech arrived, a sweep of the area was attempted, however automatic weapons fire repeatedly halted the advance until artillery fire and air strikes could be brought to bear on the enemy force. B/1/8 fired 985 rounds that afternoon and A/3/13 fired 493 rounds. Enemy casualties in the battle were 84 NVA KIA (BC) and 1 POW. WIA SFC Partee, our platoon sergeant, received a head wound in this action that landed him in a hospital in Long Binh.

This was another version of the March 4th action. After Action Report 24: Among the most successful operations during the month were in areas controlled by the 2nd Brigade deep within the CITADEL and BOI LOI WOODS. On March 4th, Company C, 2nd Battalion, 12th Infantry, performed a combat assault only eight miles northwest of CU CHI and encountered a heavy volume of fire from a well concealed and entrenched enemy battalion. The contact was reinforced by a task force of three rifle, one mechanized, and one tank company. Three thousand rounds of supporting artillery were fired. This engagement lasted until 1800 hrs. An evening sweep of the area revealed 84 enemy dead. The combined forces established a night defensive position in the area. Early the next morning, this position received a heavy attack by fire and a subsequent ground attack. Artillery, helicopters and fighters broke the assault, and caused the enemy to leave another 74 dead on the battlefield. U.S. casualties, by contrast, were light: 11 killed, 37 wounded.

Early the following morning, A/1/5 Mech received RPG, mortar, and automatic weapons fire. Artillery fire was quickly placed on the enemy. When the forward observer and the recon sergeant were evacuated with multiple fragmentation wounds, the artillery LNO took over the missions. Artillery units fired 400 rounds of 105mm, 136 rounds of 155mm, and 62 rounds of 8" ammunition prior to 0650 hours in support of A/1/5 Mech. Total enemy losses for the contact on 4 and 5 March were 176 NVA KIA (BC).

Reflecting back, 1LT Sheehan talked about Thomas Sovey, "He had a habit, or plain bad luck, of falling into dry wells while pulling flank security. During his decent into those pits, he invariably broke his glasses. Being the suspicious sort, I finagled a spare pair of glasses for him from our BN medics and carried them for several weeks before the barrage that I'll mention later. I particularly remember that Sovey coveted a Bronze Star, because I think that his dad earned it during Korea or WW2. Anyways, he was injured when a piece of shrapnel came through the ammo box-end shutters to his bunker on FSB Pershing one night when a 120mm mortar barrage landed in our platoon area. A pretty intense night! Toto and others were also wounded that night, and I think the monkey tree (bamboo thicket) took at least one hit that night, too. Sovey hopped onto one of the evac helicopters, without my knowledge. Later that night, I got my ass chewed for losing track of him. My relationship with the new company commander (CPT Melvin "Lee" Morrow, from VMI) was steadily downhill after that incident. I ran into Sovey one more time while standing in line at the Forth National Bank (curious name) at Fort Lewis, Washington in 1970. After giving him a light-hearted ass-chewing for disappearing, my wife and I gave him a ride back to his company area."

A few days later, after the March 4th battle, I had my last chance to offer some last words of advice to the guys. It was a hard time for me. Not knowing what would happen to these boys that I had spent so much precious time with. The thoughts we had, the memories of battle and laughter, all were swirling around in my head. I was overjoyed knowing I was getting out of this situation, away from death and the certainty of the unknown. I was heading home in a few more days. It was hard not to think of what fate would bring to these guys after I leave.

Years later, I wondered how I really said those goodbyes and good luck messages. Three senior NCO's were leaving the platoon at the same time, leaving behind SGT Keatts to fill the void and assuming a whole lot of responsibility. PSG John Partee was in the hospital with a head wound and shortly after, I left to go home, LT Sheehan was involved in an accident when the truck he was riding in from Cu Chi back to FSB Pershing hit a mine and he was severely burned. He ended up being sent to Japan and then home. He would return to Vietnam in a few years as an ARVN advisor. The platoon was left without any combat leadership except for Keatts, but there were guys who would be asked to step up.

Price, Hill, and I arrived in Long Binh and checked in, receiving our papers and flight orders for home. We had a bit of time available so we went over to 3rd Field Hospital to check in on Partee. He was sitting upright in bed, but unconscious. I thought he had been hit by the enemy, but Dennis Buckley says he was accidentally hit by a grenade thrown by one of our own troops and he was just plain unlucky. The grenade hit a tree branch and bounced back in the direction of Partee. He would die of these wounds on March 29th. I couldn't even say goodbye to my PSG whom I had served with since July. Besides Partee, SGT Richard Deimler from Hummelstown, Pennsylvania was also killed outright in the March 4th action. He was a friend of Elmer Lightner.

March 10th arrived and we headed to the airport with our flight orders and boarding number. We stood on the tarmac watching the plane taxi to the boarding area, shut down its engines and then waited while the ground crew rolled the staircase over to the main cabin door. Minutes later, the stairs were filled with replacements working their way down the gangway and onto Vietnam turf. They looked just like we looked arriving here one year earlier. They could not imagine yet what was in store for themselves and what was going to happen to them. As they slowly walked by those of us standing there in our clean fatigues, golden brown from the sun, somber, yet inside, our guts were churning with anticipation about getting on board and leaving this place.

As they continued to stream by, in our side vision, we could see the ground crew refueling the plane. Once the plane was emptied and we had our last shots at words of whatever came to mind with the replacements, it was our time to board. A sergeant began to call out boarding numbers and as he did, whoever held that number sounded off, showed the NCO his orders then climbed aboard the *Freedom Bird*, a Boeing 707 bound for Travis Air Force Base in Fairfield, California. It was dead silent on the plane until the wheels were up after takeoff, then a large roar was heard in the aircraft. For those of us with a window seat, we peered out downward and silently watched the landscape slip away out of sight. We were free.

Nine days later, on March 19th, Charlie Co received small arms, automatic weapons, and mortar fire on a sweep near the village of Sanh Ho a place that always gave us trouble. The artillery supporting them, B/1/8th and A/3/13th, fired 103 rounds. Results of the action were 32 VC/NVA KIA (BC) and 30 VC/NVA (poss.). In this fight which started around midnight, the VC began to attack from the tree line firing small arms and RPG's. First squad with Bruce Solem, Pat Flood, John Mikita, and Bob Walton and second squad with Billy Bowden, Jim McInvale, Vernon Becker, Percy Miller, Ron Jochum, and Allen was anchoring the right flank of the laager site as it faced a heavy wood line about fifty meters from their position. They were under a heavy assault.

The VC had managed to cut the wires to the Claymore mines. The small arms fire was so intense that they could not look up above the berm they were crouched behind. They tried to raise their weapons above their heads and fired blindly at the enemy. Richard (nicknamed "Satch") Satterthwaite and Dennis Buckley, manning the M60 machinegun were crouched behind and along a rice berm along with the rest of their squad and to the left of second squad. The rest of the company was in a square defensive position spread out over a rice paddy behind them.

The 1/5th Mech was off to their right holding their own position. Satch and Buckley could see the enemy advancing on their buddies on their right flank and exposed themselves to enemy fire while they tried to repel the assault. In the exchange of fire Satterswaithe was killed. He was a close friend to Dennis Buckley and was taken out just a short distance away. Dennis took a severe hit to his right side by the same RPG and lost consciousness. When he came to, things were just a blur. He was Medevac'd to Cu Chi to 12th Evac then later to a hospital in Bien Hoa. Shortly thereafter, he was taken to Japan where he underwent a series of operations and remained there for the next year and a half. After that, he was transported back to the U.S. for more follow up treatment.

He has had a rough time and in 1974 was retired from the Army due to his medical condition. Dennis would wear out a path to the V.A. hospital in San Francisco over the coming decades seeking medical treatment due to his injuries. SP4 Loren Jones from Unaka, North Carolina was also killed. The following day March 20th. B/1/8 Artillery fired 251 rounds and A/3/13th fired 165 rounds in support of CPT Charles Boyd's A Co who had come under attack at 0218 hrs. C/1/5th Mech also received mortar and small arms fire. Three batteries, A/3/13th, B/1/8th, and D/3/13th Artillery responded with 396 rounds of high explosive and 19 rounds of illumination. Results of the action were 26 NVA KIA (BC). SP4 Bruce Solem later commented that if it wasn't for Satch and Buckley, they would have been overrun in the firefight.

The two day period, 17th to 19th March, produced several ill-conceived enemy attempts to defeat 2nd Brigade units. A night attack on A Co 1st Battalion, 5th Infantry (Mechanized) resulted in 30 enemy dead on the 17th. Later that same day, C Co, 2nd Battalion, 27th Infantry, fought an engagement (XT565227) with an enemy force in an open area. Without reinforcement, through classic fire and maneuver, C Co killed 81 of the enemy, at a cost of one wounded U.S. soldier. That same enemy force attacked C Co two days later in a night defensive position in the same area. The cover of darkness did not improve their situation. They lost 33 KIA again. The enemy became increasingly reckless as he attacked a night defensive position of A Co, 2nd Battalion, 12th Infantry in the same area on March 18th. The massed defensive fire around the position produced another 22 enemy dead. A similar night attack on March 19th cost the enemy 26 killed when he fought C Co, 1st Battalion, 5th Infantry (Mechanized). Numerous small encounters continued in this area during the next week. On the 27th and 28th of March, the Division OPCON'd the 2nd Battalion, 27th Infantry, and the 1st Battalion, 5th Infantry and 101st Airborne to the 2nd Brigade in order to exploit the contact gained with the enemy in the area. During the month of March, the 2nd Brigade delivered a major defeat to the enemy units massed along the Saigon River infiltration corridor.

On the 2nd of April, D Co led by CPT Paul Allen, detected movement outside of its night defensive position near TRUNG LAP (XT556219). At 0045 hrs the VC launched their assault. The company engaged the enemy with organic weapons, artillery, and helicopter gunships. The enemy, attacking, fired with automatic weapons, RPG's and mortars, but inflicted no casualties. At first light, D Co counted 49 enemy bodies.

22. DESTINATION U.S.A.

As our freedom bird slowly climbed to gain altitude, the landscape of Vietnam began to fade from view. It had been a long, hard fought year and finally we were free of the stress of combat. The cabin noise slowly diminished as everyone was caught up in their own thoughts about heading home finally, to a normal life, a safe life. We flew north and landed in Osaka, Japan for refueling. It was 0300 when we landed and we had a chance to stretch our legs before taking off again. It was damn cold there with a touch of falling snow drifting through a biting and cold wind as we walked to the terminal. While there, I bought a few wall decorations as mementos, but never used them.

Our flight took us north just below Alaska and down the coast line to Travis Air Force Base, California where we landed at 1730. Because of the International Date Line, we landed a half hour before we took off from Vietnam. After stepping off the plane and walking down the rollaway ramp we walked over and got on a bus to the Oakland Army Base and staging facility where I was fitted with a new class A uniform and processed "in country," fed a steak meal, or someone said that's what they got so I assumed I had the same. Later, after the tailor was done and I was allowed to change into my uniform, I was set free to find my way home for a thirty-day leave.

It took about five hours to get through processing at Oakland. The last piece of paper I got before I departed were my orders to report to Fort Ord on April 10th. I didn't call anyone to tell them I had arrived at home. It was already close to midnight and I was sure everyone was in bed. Several other guys and I caught a cab to San Francisco where I got a Greyhound bus ticket to Santa Rosa. I arrived in town around 0230 and immediately called my friend Bill Hougen, then my sister Kathy who came down to the bus depot to pick me up. I spent the night at her apartment then we drove up to Cloverdale to see Mom and Dad in the morning. We had a great reunion.

I had been in town for several days when Patty Pressley discovered that I was back. She was the girl I had hung around with on my leave prior to going to Vietnam. I was sitting in the living room of my parents' house when I saw her white Cadillac roaring up the street toward the house. She made a big circle turn and stopped on a dime in front of the house, got out of the car, and stood there glaring. I went out to see her, but I had no words to say. It was a very awkward moment with the two of us standing there in silence.

We had been writing back and forth for a year, with many words being written, and yet I couldn't find any words to say. Mind you, it wasn't in my mind over the course of the year that something was written or implied that would lead to any promises or expectations for either of us. The signatures at the end of letters turned from, yours truly, or always, to sometimes with love. But what did that mean? We never dated, let alone kissed. And now, I'm facing an outraged woman who was taking no prisoners. We never spoke again after that day. It's one moment I regretted in hindsight as I could have been less callous, but somehow I wasn't ready for hugs and kisses. In hindsight, I hope she was able to find that special someone who she could share a life with. Years later, I felt like a heal in how I dealt with that moment in time and how I reacted to her.

After my leave, I reported for duty to Fort Ord, Monterey, California. My orders had me reporting to a training company, but when I knocked on the door of the First Sergeant and showed him my orders, he had a cow, swearing and such, then told me I wasn't needed there and to go to HQ and get reassigned. So, I went searching for HQ, found it and got new orders issued. I drove over to my new assignment, reported in and was told I would be reporting to S.E.E. Committee. AITCG was a training company of instructors for advanced individual training. My particular committee group handled three different areas of training. We taught S.E.E., which stood for survival, escape and evasion, of which I was the PI (principal instructor). I also taught a class on the M113 APC and the committee ran the grenade range.

Being that I was a Staff Sergeant, I had to drive down to the range early in the morning before the trainees arrived and count the lots of munitions (grenades) and sign for them with the ordnance boys. Duty at Fort Ord was great and was over before I knew it. Rank had it privileges and being a SSG, I did not have to have a pass to leave the post. I came and went as I pleased. I was three hours from home and would drive home every weekend after training was completed, then drive back to Fort Ord in time to be at reveille at 0600 Monday morning. While there, I was also the barracks sergeant and had a private room which I shared with another NCO. I was responsible for reveille and accounting for the troops every morning. I got out of the service on September 6th, 1969. Since I was a combat veteran, I did not have to fulfill any further obligation for reserve duty. I received my Good Conduct Medal as I left Fort Ord and later my Honorable Discharge papers.

As I returned to civilian life, no one questioned where I had been or what I had been doing. The conversation about Vietnam, if there was one, continued to be negative in most circles. I don't recall saying much if anything to my family nor did they ask any questions about my service. Most Vietnam veterans initially feel some degree of isolation or remoteness from society. For some, that feeling and state of mind will never leave them. They have difficulty in coping with their battlefield experiences and the emotions that are tied to those memories. All that is pent up inside and there's no relief from all the thoughts and flash backs that are swirling around in their head. Driving around town, it was hard to understand the "normalcy of civilian life and the carefree attitude" that you could observe. It was such a stark contrast to packing a rifle every day and dealing with firefights and death.

I went back to work at Pacific Bell. They held a position for me or created one, upon my return and release from active duty. I held numerous labor jobs, i.e. supply man, lineman, installer, PBX installer, over the years, then I was promoted to supervision where I ended my career as a second level area manager working in Business Services, leaving the company in 2000 after 33 years. My wife Jeanne and I were married in 1973 and we had two sons, Matthew and Jesse. Jeanne also worked for Pacific Bell and retired as a technician shortly after I did. She was very supportive in my quest to reunite with my fellow veterans and has enjoyed getting to know some of them and their wives over the years.

Vietnam was part of my past and remained so but was never out of my conscious mind. It was not until the late 1990's when I began to look for men from my company and platoon. I was sure who made it home from the service period I was there because I checked the names on the Vietnam Wall. At first there were infrequent phone call attempts to locate guys and responses to email addresses left on military websites. This went on for another ten years or more. The only person I found and communicated during this time was Elmer Lightner and it took six years of emailing back and forth before one of us picked up the phone and called. Around 2009, I became dedicated to pursing this idea of finding guys and since then I have made contact with dozens of former Charlie Co veterans and we have gotten together, holding many reunions. Many of the officers who served in Vietnam who stayed in the service until retirement rarely rose above the rank of Lieutenant Colonel. That could be for many reasons not of their making. For example the military always downsizes after ending conflicts and that caused the services to thin the ranks of senior officers at times. The process then needed to be reversed as early exits or retirements put a strain on the needs of the military. The number of allotted positions in each rank also depends on the speed or opportunity to be put on the "list" of potential promotions.

There were a few exceptions who made Brigadier General or higher, as was the case for battalion commander R. Dean Tice who achieved three stars, Lieutenant General upon retirement, and COL Marshall Garth, the brigade commander who retired as a Major General.

I learned that soon after I left, my squad split up. George Toto extended to keep his brother out of Vietnam and after being wounded three times, went to Leadership Development School for newly promoted sergeants as an instructor, along with Terry Corum. Jesse Tostado was sent to Okinawa where he did bomb ordnance removal. Ed Wales left the field and went to resupply. Curtis Tassen would make sergeant and take over the squad later. Russell Zimmerman finished his tour in the field.

The sad note in looking for members of the platoon was the discovery that some men had passed away at a young age. Why? Many deaths had been attributed to our exposure to Agent Orange, a defoliant used in Vietnam to eliminate the hiding places of the enemy. This chemical was sprayed by airplane onto the areas we were operating in. The most heavily sprayed areas were in III Corp, the area west of the capital city of Saigon. It took many years and ultimately a decision by the Supreme Court forcing the U.S. government to take ownership of the health issues created by the usage of this chemical. There are 15 presumptive diseases currently listed. If you are diagnosed with any of these you may be entitled to compensation and treatment. The Veterans Administration is currently studying an additional three to be added to the list. Better than 60% of combat veterans have been diagnosed with some health issues, including PTSD. I am part of that list.

In 2011, I launched my website, www.212warriors.com, dedicated to the entire battalion, preserving our history of our participation in the Vietnam War. As a result of this website, I was invited to Fort Carson, Colorado Springs, Colorado where the 2/12th is on active duty. I have had the honor of being the guest speaker before the entire battalion twice and have participated in numerous change of command ceremonies for both the battalion commanders and the CSM's, command sergeants major's. The unit has been great and extremely gracious to former " Warriors" and we are welcomed anytime we visit the post. The officers continually emphasize that the "door is always open to us." It has been a very emotional and healing experience for us who have visited. The accent is always focused on us and they douse us with praise and gratitude for leading the way and what they have learned from our experiences of the past. All we can say in return is "thank you and thank you for YOUR service."

The website opened many doors for me and I have met and interacted with numerous other veterans who served in the other companies of the battalion. The website has been a gateway for vets to find their lost buddies or to find out that someone has passed on after the war was over. It has been a vehicle of healing. I have spoken to many family members who have finally received the final word on what really happened to their loved one. In one way or another it has also been a stress reliever and a semi cure for PTSD. I produce a newsletter promoting the website and try to keep the veteran audience informed of current events as well as upcoming reunions.

In 2017, I was on the 12th Infantry Regiment Chapter Association's Executive committee for the placement of a monument at the Walk of Honor at Fort Benning, Georgia. I assisted in fund raising and promoting the project plus attended the dedication ceremony on May 31, 2017. As part of this team, I was honored to be nominated as a Distinguished Member of the 12th Infantry Regiment by the Secretary of the Army.

Before I forget, one of the most important elements of this entire journey that I briefly touched on earlier has been the reuniting of many of the men I served with. I have worked hard to seek out and find members of my platoon and others who I knew in Vietnam. I used a combination of letters and phone calls as ice breakers and if it hadn't been for the advancement of the internet, many of these men I would not have been able to locate.

When I began looking for veterans I served with, seriously around 2008 and 2009, Charlie Co, the 1970-71 guys who served together, met annually in Gatlinburg, Tennessee. I managed to get a number of the guys to go to this reunion starting in 2010. George Toto and Jesse Tostado from my squad, Dennis Buckley (since has passed away), Elmer Lightner (1st Plt) and Michael Sheehan all attended. The first evening there, sitting in the bar, having an adult beverage with the gang, Mike and I had a good light hearted conversation about being our PL. He was the first officer willing to join our reunion. I wanted him to feel at ease and not out of place being with the EM's. He's been a great guy to get to know, as have all the men I have had the privilege to reconnect with. The group has grown since that first meeting.

Some vets were satisfied just to have talked once or twice on the phone. I can only be grateful for that small period of time. Others simply didn't want anything to do with any aspect that will recall their time in Vietnam and I had to respect that point of view as well. I am not in a position to be judgmental about those who have struggled with their service time or their decisions whether to reengage or ignore these invitations to get back together.

The men who I correspond frequently with reaches out beyond those I served with in Charlie Co and I have made friends with many, many men from all the company's in the battalion, primarily because of the website I built. That has been an added blessing and opportunity to encourage many to seek help from the VA who have been reluctant to do so, to encourage them to file claims with the VA, to look for buddies, to answer questions about what the battalion was doing on specific dates to help with their recall and the list goes on.

Lastly, has been the rekindling of friendships with those in my company who want to meet up every year just to share stories and see each other again. The wives all get along and it's such a wonderful experience each time we meet. There are no ego's, or personality conflicts to manage, just a great group of people. I have gotten to know them all looking through a prism of age and of understanding and we have grown together through these reunions. I know there are many other small groups that do the same thing with their platoons or companies and they should be applauded for keeping that passion of caring about each other, along with the bonds they created so long ago..

Today, most of the platoon has been found and contacted and we have tried to find a few from the other C Co platoons with limited success. The door is open for those who may want to reach out at some point and join in the reunions. There are about 12-14 who on and off, attend our annual group get together. I hope more will join in the fun.

As I close this final chapter, I am always aware of the feelings I have, having served in the military for my country. It may not have been my own choosing when I was drafted and ended up in combat. However, I will always be prideful of what I went through and every time I see the colors of our Nation's flag fluttering in the wind I get emotional, for myself and for those who have so gallantly fought and died for the RED, WHITE and BLUE.

SERVICE RECORD

DRAFTED — SEPTEMBER, 1967
DEPARTED ARMY — SEPTEMBER 1969

Basic Training: Fort Lewis, Washington
AIT: Fort Polk, Louisiana
Served in RVN — March 1968 to March 1969
Rifleman and ammo bearer for MG crew
RTO for Platoon Leader
Squad Leader
Platoon Sergeant

AWARDS

Combat Infantry Badge
Bronze Star Medal
Purple Heart W/OLC
Vietnam Service Medal with Silver Star
Republic of Vietnam Campaign Medal
Good Conduct Medal
National Defense Medal
Republic of Vietnam Cross of Gallantry with Palm Unit Citation
and Individual Award

Republic of Vietnam Civil Action Honor Medal, First Class

Returning to the U.S. I closed out my military service at:

Fort Ord, California, AIT Committee Group as a principle instructor in S.E.E. (Survival, Escape and Evasion), Armored Personnel Carrier and assisted with the hand grenade range.

EPILOGUE

STATISTICS AND WHAT HAPPENED
AFTER THE WAR ENDED IN 1975

Because of faulty record keeping, the number of veterans who actually served in Vietnam or in the waters around Vietnam was estimated to be between 2,709,918 and 3,173,845 of which 25% were drafted. This number is often confused with the number of military veterans who served during the Vietnam War Era and served everywhere BUT Vietnam. A total somewhere between 9 to 12 million people claim to have served in Vietnam, as reported by various veterans groups and is reflected somewhat in the last Federal Census. So obviously, some people are not being honest about serving in the military at all, let alone being in Vietnam. As of this writing, the numbers of Vietnam veterans is dwindling and the American War Library estimates that approximately 610,000 Americans who served on the land or in the air over Vietnam between 1954 and 1975 are alive today and approximately 164,000 who served at sea in Vietnam waters are alive today. (Source: American War Library)

The average infantryman in WWII saw 40 days of combat over a four-year period. The average infantryman in Vietnam saw 240 days of combat in one year. In 1968, there were 62,800 to 67,700 combat troops, the rest were support troops. The 25th Infantry Division suffered 4,561 KIA's for the entire war, second to the 1st Air Cav which had 5,464 KIA's. From 1968 to 1969, there were a total of 28,679 KIA's, over half of the total combat forces killed in the conflict. California suffered the greatest losses in the war with 5,585 casualties, New York was second with 4,119 loses. Out of the officially listed number of casualties of the Vietnam War, currently at 58,286, 40,934 were killed in action and 5,299 later died of wounds. There were 236 homicides and 382 self-inflicted deaths. It was a conflict and not a war since Congress never officially declared war on North Vietnam. However, like Korea, everyone refers to it as a war. The average age of the soldier was 22.8 years compared to 26 years for the WWII veteran.

All U.S. combat troops were pulled out of Vietnam by 1973, two months after the signing of the Vietnam peace agreement between the U.S. and North Vietnam in Paris, France. February 12, 1973, Operation Homecoming began the release of 591 American POW's from Hanoi. By March 29, 1973, the last remaining American troops withdrew from Vietnam as President Richard Nixon declared, "The day we have worked and prayed for has finally come."

What remained in country were about 7,000 U.S. Dept. of Defense civilian employees. The feeling back in Washington was to 'cut our losses and move on, which they did with the amendment Congress passed in June, 1973 forbidding any further military efforts in Southwest Asia.

As soon as the U.S. combat presence was gone, and with the knowledge that the U.S. had officially ended support to South Vietnam, the North Vietnam Army accelerated its plans to consume and lay waste to any remnants of the ARVN army and its government, thus violating the cease-fire achieved through the Paris Peace Accords. By the end of 1974, the South Vietnamese authorities reported over 80,000 soldier and civilian losses, making it the most costly year of the war. On April 30, 1975, the fall and capture of Saigon brought an end to the South Vietnamese government and its army and an end to the Vietnam War. The Vietnam War, also called the Second Indochina War and known in Vietnam as the Resistance War Against America (Khang Chien Chong My) or simply the American War. The war extracted a huge loss in human life. Estimates of the number of Vietnamese soldiers and civilians killed vary from 966,000 to 3.8 million. Some 240,000-300,000 Cambodians, 20,000-30,000 Laotians, and 58,286 U.S. service members also died in the conflict.

On July 2, 1976, North and South Vietnam were merged to form the Socialist Republic of Vietnam. As part of this process, over 300,000 people from the south were forced to attend re-education camps in the following years. Many endured torture, starvation, and disease while being forced to perform hard labor. Many from the south were fearful for their lives and chose to escape the country by boat. As many as 200,000-400,000 boat people lost their lives and died at sea trying to escape from the heavy-handed Communist government.

In Phnom Penh, the capital of Cambodia, the Khmer Rouge took over. In subsequent years they would eventually kill one to three million Cambodians out of a population of eight million. It wasn't until 1978 that Vietnam invaded Cambodia and ousted the Khmer Rouge, who was supported by China. This led to a border conflict with China in 1979. From 1978 to 1979, some 450,000 ethnic Chinese left Vietnam by boat as refugees or were expelled. The impact of the Khmer Rouge rule contributed to a 1979 famine in Cambodia, during which an additional 300,000 Cambodians perished. It is estimated that over three million people left this region as refugees. Between 1976 and

1998, an estimated 1.2 million refugees from Vietnam resettled in the United States. (Vietnam War—Wikipedia)

Initially after the war, the south was shunned by the north for almost 10 years. They did nothing to aid or assist the population. Slowly this began to change. The communist government could see what was needed to survive as a country. It needed to be a nation state willing to trade in the free markets of the world. The government also realized it needed the rice production of the south to succeed.

Slowly the government lifted the sanctions that were in place. The people were granted the privilege to own property and to produce a crop that could be sold for individual profit, not just to support the government itself. The people through their constitution were granted free education; however, even today it does not quite work like that, as parents have to pay to send their children to school. A citizen was free to move about the country but must first apply to the government to receive its approval or you cannot relocate.

Pay is still suppressed and wages even with good jobs like teachers and nurses lags behind other countries of the region like Thailand. The cost to own a home and land is out of reach for many. Farmer coops exist to share in the cost of machinery to plant and harvest the rice crops. Capitalism is on the rise in this country and it makes for a happy population.

The biggest change was in housing construction. Gone is the mud hut with the thatched roofs or for some maybe corrugated roofs. All the homes are brick with sturdy roofs. Most still lack doors and windows, but in that climate, it's just not needed. The square footage is small, maybe 600-800 square feet. Every home today has electricity which was accomplished in the 1990's and most has some form of communications. It was interesting to drive by homes and see only one light on and sometimes a TV, but not the entire house would be lit. The production of electricity is still a major problem for the country. They can't produce hydroelectric, coal does not exist, and nuclear did not appear to be an option either.

Cell phone service is available everywhere in the country. Most roads are now paved with asphalt and the cities are modernizing with new bridges and roads. We found only one unpaved road and we had to go out of our way to find it. It still looked the same, the bright red earth and plenty of potholes. The country is still not up to the same standard of modern countries, but they are moving forward.

Manufacturing of cars and trucks, clothing, shoes, and other products besides the staple of rice production helps their economy. They are trying to grow their tourist trade too with hotels, golf courses, and even some gambling. It is an exciting country to visit as a tourist and is hard to beat the beauty of the place. In November 2010, Ed Wales, Dennis Buckley and I returned to Vietnam and visited the area around Hue down to Da Nang before returning to Dau Tieng, Hoc Mon, Trang Bang market and 6 Alpha, then FSB Pershing. We walked the lands we once patrolled and fought in and visited the sites of some of our firefights, placing flowers in honor of our fallen comrades. Traveling about the country in 2010, we noticed a complete absence of authority outside of an occasional policeman. Even in downtown Saigon, you could only find traffic control officers.

Driving around the country side you could easily identify government buildings as they were all painted in a bright yellow with some red trim. The schools, hospitals, and military bases all looked the same. The schools were teaming with children and at the end of the day when school let out, the parents were there to greet and take their children home, mostly using their motorcycles. It was fun to see women in high heels and dresses working their way through traffic on their bikes. It was even more amazing to see what they could stack and carry on one of those vehicles. I saw as many as five people on a motorcycle. Delivery men would balance stacks of goods on the front and back racks while holding another package on their lap and never strap anything down.

Hanoi remains the capital of the country. Saigon, now called Ho Chi Minh City, is home to over nine million residents and has about five million motorcycles clogging the streets. Very few traffic lights exist in the city and there is no such thing as a four way stop. Traffic just merges from all directions, with an occasional polite toot of a horn to keep someone moving. Everyone is patient and polite, whether driving or walking. The people were kind and never hinted about the past and the war. For many, it isn't even a memory due to their age. In 1968, there were 41 million in the country and today that number is over 95 million. The city is growing and replacing the slums of the west side of the city. They are even constructing a light rail system. It's a great country to visit.

If you travel there, visit the area around Hue as there is great cultural and historical exposure to take in. The food is excellent everywhere you travel, sit-down restaurant or street vendor, try both. The people are young and friendly and show favor toward Americans. The beauty of the country is unmatched and still has the feel away from the city that time moves more slowly in the countryside. I thoroughly enjoyed my time traveling back to Vietnam and seeing the country from a different point of view.

GLOSSARY

NOTE: Unit designations can be tricky because they use a combination of slashes and at times dashes to define what battalion and regiment they belong to. This is true of infantry, cavalry and armored units. It can also be true of a unit that was assigned to another organization, but was only a part of that unit, not the entire unit as in the case of armor units. A regiment could be broken up into battalions and each battalion assigned to a different division or brigade as a support element, hence the "-" means the unit is separated from its parent organization . But for the sake of this writing, we follow the usage as it applied to the Vietnam War. It's too complicated for this book to outline all the military applications and changes adopted in today's military. I hope what you find in the glossary will carry you through.

(-26) — Any reference to a company with this notation indicates it's not at full company strength, but missing an element or platoon. This example the 2nd Plt is not with the company.

2/12th — Numbers identify specific units with the first number (2) indicating the battalion. The second number (12) is the regiment. Type of regiments are infantry (Inf) cavalry (Cav) and armored (ARC).

2/12th B Co (-36)—Indicates B Co minus 3rd Plt.

2-34—This symbol is defining a "battalion and armor regiment because the "-" symbol is being used.

AHC—Assault Helicopter Companies, aerial lifts or transport units supporting combat operations for ground infantry equipped with Huey UH-1D's.

Airlifts—Air operations usually coordinated by the BN S-3 Air Operations officer. Number of aircraft used in a "lift" is based on availability and the size or number of troops needed to be relocated to a landing zone.

AIT—Advanced individual training. Follows BCT or basic combat training; both last eight weeks. AIT teaches you how to perform your MOS (Military Occupation Specialty), a combo numeric alpha code that identifies your occupation. Infantry is 11B—the next two digits that follows represents your rank. 11B10 is a E3 or Private; 11B20 is a E4 or SP4 (Specialist 4th class); 11B40 is a E5, SGT or E6, SSG etc.

AO—Area of operation, or TAO, Tactical Area of Operations; physical area assigned to a unit to control

AP—Ambush patrol

APC—Armored personnel carriers, or M113's, usually referred to as *tracks*

ARCOM—Army Commendation Medal awarded either for merit or valor

Armor(ed) — Units using heavy tanks; Armored Calvary Regiment (ACR) units use light APC's for reconnaissance; 3-4 platoons per troop (Co), 2-4 troops per squadron (Bn) — shown as 2-34th (with a dash not slash) 2-34th is 2nd Squadron, 34th Armored Regiment.

Article 15, Uniform Code of Military Justice (UCMJ) — Military punishment for a minor offense not requiring a judicial hearing handled by the commanding officer who can impose for example, forfeiture of pay, reduction of rank, extra duty or reassignment if found guilty. Usually done for misconduct offenses.

Artillery unit — Varied by caliber of guns each battery was equipped with. Usually a Battery (see Battery) was assigned to a specific FSB or basecamp. Shown as A Battery, 1st Battalion, 8th FA (Field Artillery) i.e. A 1/8th FA

ARTY or DIVARTY — Artillery or division artillery.

ARVN — Army of the Republic of Vietnam, also referred to, or called South Vietnam.

Avn Bde or 25th Aviation Brigade — Organic to the 25th ID. Consisting of one squadron of Huey UH-1B gunships (Diamondheads) and one squadron of Huey UH-1D troop transport ships (Little Bears), stationed in Cu Chi.

AW — (automatic weapons) Autoloading rifles capable of full automatic fire.

Azimuth — Direction of travel using a compass and stated in degrees (i.e. east is 90 degrees)

Battery—A force of 100+ men (company size), six guns or howitzers with crews and support elements led by a Captain. A battery is equipped with one of these type howitzers: 105mm, 155mm, 175mm or 8 inch.

BC—Body count

BDE—Brigade or Bde (abbrev). Consists of 3-5 battalions; combination of ground troops, mech., armor and artillery

Beehive round—An anti-personnel round, also called a flechette round (filled with metal darts), direct fired from artillery guns against enemy troops at ground level.

Big Boy—Nickname for M48 Patton tanks.

Bloused boots—To tuck the fatigue pants into the boots so the pant bottoms don't show.

BN—Battalion or Bn (abbrev). Usually consists of four companies of 100–160 men each

Boo-ku—Vietnamese for many, lots, or a long distance.

Boonies—Term to describe being out in the jungle or outside and away from basecamps, FSB's or the city. A quick boonie pit stop can mean a "bathroom break."

Brigades—Combat maneuvering elements of the 25th Infantry Division during 1968. This configuration changed from month to month depending on strategic needs at the time. Each brigade was assigned a specific TAO, tactical area of operation with geographic boundaries. This is a listing of the maneuvering battalions in 1968 and their nicknames:

1st *"Lancer"* **Brigade: 4/23rd (Mech)** *Tomahawks,* **2/14th Inf** *Golden Dragons,* **4/9th Inf** *Manchus,* **and 2-34th Armor** *Dreadnaughts*;

2nd *"Flame"* **Brigade: 1/5th (Mech)** *Bobcats,* **1/27th and 2/27th Inf** *Wolfhounds*;

3rd *"Bronco"* **Brigade: 2/22nd (Mech)** *Triple Deuce,* **3/22th Inf** *Regulars,* **2/12th Inf** *White Warriors*

Brigadier General—BG, one star and usually the Assistant Division Commander or staff.

Bug Juice—Insect repellant

C4—A white plastic explosive, one pound in size, and capable of generating severe force. A small block about 1" by 2" equaled a M1 grenade. Most G.I.'s carried a small block in their pockets. They would break off a small piece the size of a marble, light it, and it would burn like Sterno and was used to heat c-rations.

CA—Combat assault, generally the deployment of an infantry unit by helicopters

Cal—Caliber, used after a number, .51 cal, to indicate the bore size of a weapon's barrel.

Call Signs—Names used to identify units and command heads when chatting on the radio networks.

Can Cuoc or Can Cuoc Cong Dan—Vietnamese ID Card

Cav or Cavalry Troop—Configured as three infantry platoons mounted on APC's and one airmobile platoon equipped with Huey's performing reconnaissance intel and troop ground support via gunships. A "troop" equals a company in strength. Three troops make up a squadron.

Charlie—Nickname for Viet Cong (VC guerrilla fighters) or NVA soldiers.

Chicom—Slang for Chinese Communist and used to describe some of the Chinese military hardware the North Vietnamese government used.

Chinook—Boeing CH-47 helicopter, a twin rotor aircraft used for transporting 25-30 troops and can be used to lift heavy objects using cargo nets or slings attached to a hook underneath the aircraft.

Chopper—Nickname for helicopters, also called "slicks or Hueys."

Claymore mine— A curved shaped device with foldable legs, plastic in construction that housed a steel plate which broke up into fragments by the detonation of the one pound of C4 explosive that was encased behind it.

Click—A click is equal to 1000 meters, the standard measurement used on the topographical maps. Each square on our maps were 1000 x 1000 meters. The same measurement was used by artillery.

CMAC—Capital Military Assistance Command, responsible for protection of Saigon and immediate area.

CMDR—Commander and is used to reference a battalion (BN CMDR) or brigade commander (BDE CMDR).

C.O.—Commanding Officer

Co —Company, 100-160 men; four plts, three rifle, one heavy weapons (81mm or 4.2 inch mortar).

Compass headings—North, south, east, west; northwest (NW), northeast (NE), southwest (SW), southeast (SE)

CP—Check point # given to a prominent point such as a geographical reference, i.e. crossroad, village, ruins, etc., used to establish one's location on a map via radio traffic.

CS (gas)—A type of irritant tear gas or riot gas, white in color that cannot be breathed, causes coughing, uncontrollable shutting of the eyes plus tears.

CS Ship—Special helicopter (Huey) equipped to spray CS Gas from the air.

Det Cord—Detonation Cord, a white explosive powder encapsulated in a plastic cord about 5/16th in thickness, filled with explosive powder, that is used to set off plastic shape charges simultaneously.

It is set off by a combination of shock and heat, usually a blasting cap.

Didi Mao—Vietnamese for "go, leave, get out of here."

Donut Dollies—American Red Cross workers who would, on occasion entertain and visit the troops in the field. Nicknamed for serving coffee and donuts in WWII. They were volunteers.

D.O.W.—Died of wounds

DT—Short for Dau Tieng, basecamp for Camp Rainier and 3rd Brigade HQ (HHC), 25th ID.

Dustoff—Term for air ambulance service via UH-1 helicopter, equipped with a medic and used to transport wounded to hospitals. Also called a Medevac.

Eagle Flight—A helicopter ride provided by a AHC, usually anywhere between 6-12 Hueys for the purpose of conducting combat assault operations.

EM—Enlisted men. Includes RA (regular army, men who enlisted) and U.S. (conscripted—draftees). Your dogtag (metal ID) has name, your serial no. that starts with RA or U.S., religion, blood type.

FA—Field Artillery, also called DIVARTY for division artillery; operates howitzers in sizes of 105mm, 155mm, 175mm and 8-inch. Components of a FA unit are a battalion comprised of three batteries of six guns each.

FAC—Forward Air Controller, a spotter for TAC, tactical air command (jets) who usually operates a Cessna O-1 Bird Dog, or a split tail North American OV-10 Bronco—See TAC for more info.

Fast Mover—Fighter jets used for close in ground support.

FDC—Fire Direction Control, the unit which coordinates and directs mortar and artillery fire support between the forward observer and the gun crews.

Fifty—Short for the .50 cal MG.

Firefight—Exchange of small arms fire with enemy combatants.

F.O.—Forward observer, usually an officer from the artillery unit that is supporting the ground operation. He calls for, controls, directs, and specifies number and type of artillery rounds to be used.

Foo Gas—Jellied gas stored in a large drum up to 55 gallons in size. Set off by a shape charge of plastic explosive and used as a defensive, last resort threat against ground attacks.

Four Deuce—Refers to the M30 4.2 inch heavy mortar which had a range of 6,840 meters.

FSB—Fire Support Base, a laager site protected by a company or battalion of infantry who in turn is providing security to an artillery battery.

Gooks—Slang for any enemy (VC) or civilian seen that can't be identified as hostile or friendly.

Grenade Launcher—Refers to a variety of weapons capable of launching; RPG's and 40mm grenades by individuals or some which are mounted on helicopters (LFT's) or fixed wing aircraft.

Grenadier—G.I. quipped with a M79 40mm grenade launcher, a single shot breach loading gun which shoots 40mm HE, WP and anti-personnel rounds up to a distance of 400 meters.

Gun—Refers to artillery pieces—see Artillery units

Gunship—Nickname for armored UH-1 and AH-1 helicopters used for ground support.—See LFT

HE—High explosive, describing the type of ordnance being used by artillery or mortar fire.

Helicopters—We mostly used Bell Iroquois "Huey" UH-1D's also called choppers, slicks, birds for air transportation of infantry

HHC—Headquarters, Headquarters Company. Generally refers to the command group supporting battalion, brigade and division levels. For the Bn Cmdr, it consists of the X.O. Executive officer (2nd in command), the officer staff, designated by S-1, Adjutant Officer, S-2, Intelligence, S-3, Operations, S-4, Logistics, S-5, Civil Affairs officer. Included to as part of this command, is the Chaplin, M.D. BN surgeon and the medical section, Communications, Heavy Weapons 4.2 inch mortar platoon and Recon Platoon. Some of this was reconfigured in 1969 and became part of Echo Co Each S-* officer had an assistant to aid in planning and execution.

Hooch—Slang for building of any type used for sleeping and living quarters including civilian grass huts.

Horn—The handset for a PRC-25 short distance radio used by the field forces.

HQ—Usually refers to the company commander and his staff, 1st Sergeant and clerical staff.

HRS or hrs—Hours

Huey—Nickname for a Bell UH1 type helicopter.

Humping—No, it means walking.

ID—Infantry Division

IN or INF—Infantry, also shown as 2/12th in describing an infantry unit. The first digit is the battalion number, followed by a slash and the second number, the regiment number.

KIA—Killed in action—also referred to as a "Kilo."

Killer Junior—Timed fusing set on 105mm High Explosive (HE) rounds designed to air burst just above ground level and used to defend ground attacks. Invented by the 1/8th Artillery.

Laager—See NL

LAW—Light Anti-tank Weapon, similar to the VC's RPG, but is a use once and toss weapon. Not as explosive as the RPG.

Legs—G.I.'s or ground infantry.

LFT—Light Fire Team, usually a pair of Bell UH-1C helicopter gunships equipped with 2.75 inch rocket pods and a minigun.

LIB—Light Infantry Brigade

Lift—See Airlifts

LMG—Light machine gun

LP—Listening post, two or three men are located several hundred meters out from a laager or FSB site to provide an early warning for the main force.

LRRP—Long Range Reconnaissance Patrol; a team of 6-10 men who operate behind enemy lines collecting intelligence. Changed from F Co 51st Infantry to F Co 75th Rangers.

LTG—Lieutenant General, three stars—normal rank of a Corp commander.

LZ—Landing zone. Area designated to land aircraft to drop troops off during combat assault or aerial operations.

Map Grid Coordinates—Topographical maps are divided into grids; each is 1000 meters square. Each grid is divided into 100 squares, 100 m by 100 m, 10 across and 10 up. XT is one the large maps for our area.

MAT—Mobile Advisory Team who trained Popular and Regional Forces on tactics.

Mech or (M)—Short for mechanized; a mobile infantry company equipped with APC's or M113 armored personnel carriers. Shown as 2/22 (M) or 2/22 Mech. Each brigade was equipped with a mechanized company.

MEDCAP—Medical Civil Action Program—provides field medical services to local villages such as vaccines, medical treatments, dental work, wound care etc.

Medevac—See Dustoff and Chapter 18

MG—Machine Gun or M60 for short.

MG—Major General, 2 star, normal rank of a division commander

Military Phonetic Alphabet—Specific words used in substituting letters of the alphabet so there is no confusion to what is being said in communications. For example, Alpha for A, Bravo for B, Charlie for C...Lima for L, Papa for P, Zulu for Z etc. Google search for full alphabet.

Minigun—M134, also called a Gatling gun, six barreled rotary machine gun capable of firing 2,000 to 6,000 rounds a minute using 7.62mm ammunition. On aircraft they could be larger in scale, firing 20mm cannon rounds and known as a M61 Vulcan.

Mortars—A short smoothbore gun for firing shells at high angles. The U.S. used 60mm, 81mm and 4.2 inch mortars; VC/NVA used

60mm, 82mm and 120mm. Distances varied by mortar size; 70-5700 meters.

MSR—Main supply route, usually referred to Hwy 1 or QL-1.

NCO—Non-Commissioned Officer who holds the rank of Corporal and 5 levels of Sergeants, E5 to E9 and has the authority to command personnel.

NCOIC—Non-Commissioned Officer in Charge.

NL—Night laager, a circular defensive position consisting of foxholes or bunkers dug in the ground and usually concertina wire strung around the outside of the perimeter of the bunkers, and occupied by several platoons or a company size force.

NLF—National Liberation Front, also known as the Viet Cong (VC), was an armed communist political revolutionary organization in South Vietnam; a military force which fought against American forces.

NVA—North Vietnamese Army, regular army of North Vietnam, dressed in military uniforms, unlike guerrilla or VC forces which generally wore black pajama clothing.

Numba One or Numba Ten—Vietnamese kids used terms to mean one is great and ten is bad.

On Station—See Station

OPCON'd—A term describing a unit, platoon, company etc., that is temporarily assigned to another unit outside of their direct chain of command, i.e., battalion, brigade or division, meaning "operational controlled" by another command for the purpose of carrying out a mission or operation. Once the assigned is met, the unit is released from being OPCON'd and returns to the control of its original officers for further orders.

Organic—Items such as units, equipment or weapons that are a staple or permanent to the parent unit i.e. specific units in a brigade or division makeup.

PB—Patrol Base. A location created with some defensive capability in which a patrol may operate out of temporarily and usually left unprotected when not occupied.

PF—Popular forces, a form of militia supporting RVN or South Vietnam.

PIT—A term to describe a mortar or gun/howitzer emplacement, usually protected by a 360 degree sandbag wall.

PL—Platoon leader, a first or second lieutenant, commanding 30-40 men.

PLT or Plt—Platoon—four squads, three rifle, one heavy weapons (2 MG's) 6-10 men each.

PSG or PS—Platoon sergeant, usually an E6 or E7.

Puff the Magic Dragon—See Spooky or Spectre

Push—Slang for radio frequency channel used by a given unit.

PZ—Pickup zone. Area designated to land aircraft to pick up troops after conducting ground operations.

Recon—A small party, squad or platoon sent to sweep an area to collect intel, much like a "sweep" operation.

Revetments—A barricade protecting aircraft from blasts.

RIF—Reconnaissance In Force, more than two platoons, or larger force, sent to sweep an area, collecting intel and prepared to engage the enemy.

RPG—Rocket propelled grenade, designed for use against armored vehicles but deadly as well against infantry.

RTO—Radio telephone operator. Operates a short-wave solid-state portable FM radio (AN/PRC-25) which weighs 20 lb and could transmit between 3-7 miles depending on terrain and length of antenna used.

S-1, S-2, S-3 etc—See HHC for details

SA—Small arms—These are single shot rifles or pistols.

Sapper—Nickname for a military or combat engineer

Search and Destroy or S&D—Find the enemy and eliminate them. Intel would be a strong indicator that finding the enemy was highly probable and likely.

S.E.E.—Survival, escape and evasion. A class taught in AIT training.

How to avoid the enemy, navigate and survive on your own behind enemy lines or if separated from your unit.

Short or Short-Timer—Someone who's reached 30 days before going home to U.S.

Short Hop—Slang for catching a ride on an aircraft to move from one location to another.

SIT REP—Situation report. Reporting your status to command or staff, your location, current activity, possible enemy action if engaged.

Sixty—Short for the M60 MG

Slant eyes—Slang and derogatory term for Asians in general.

Slick—Nickname for Huey helicopters used to transport troops because they are void of heavy weapons other than two door gunners on M60 MG's

Spectre—The newer version of Spooky, using an AC-130 as the aerial platform equipped with 3 Gatling guns, 20mm cannon or 105mm cannon.

Spooky—The Douglas AC-47 equipped with Gatling guns and later 20mm cannons to provide ground support. They also could drop aerial flares. Their call sign was "Spooky." Also called "Puff, the Magic Dragon."

Squadron—identifies a battalion for armored cavalry units of 42

tanks or Air Cavalry which is aerial helicopter equipped units. Abbrev as 3/4 Cav or Third squadron, Fourth Cavalry Regiment.

Station or On Station—A term meaning that air support had arrived and was circling overhead waiting for instructions or engaging the enemy.

Sweep—To patrol through an area checking for hostiles or to gather information.

TAC—Tactical Air Command, responsible for coordinating USAF ground support strikes by jet aircraft in support of ground forces and pilots in the air. A pilot, the Forward Air Controller "on station" meaning flying overhead of the action, usually flew a "Bird Dog" O-1 Cessna, O-2 Skymaster or a Rockwell OV-10 Bronco and directed the air attack.

TB—Short for village of Trang Bang.

TF—Task Force, an assembly of various units put together for a specific objective, the makeup depending on the target. It could be a blend of infantry, mechanized or heavy armor or all three.

TM A, B or C—Stands for Team A, B or C and made up of 2/22nd Mech (M) and 2/12th Infantry platoons while in Trang Bang. If it says **TM B-INF**, it means the infantry is operating without the APC's on that day.

TOC—Tactical Operations Center. The TOC serves as the battalion command and control hub, assisting the commander in synchronizing operations between field forces and support elements, i.e. air, artillery and other ground forces.

Tracks—Slang for APC's or armored vehicles including tanks.

Troop—Describes an armored "platoon" of four tanks plus 2 HQ (headquarters) tanks; can also be used to describe Air Cavalry helicopters.

Tunnel Rats—Volunteers who crawl down into the underground tunnel complexes the enemy used to hide, store provisions, set up a hospital or conduct operations in these facilities. They would search these tunnels using a knife or .45 cal pistol for protection.

Unit—Substitution for describing an army element such as squad, platoon, company or battalion that has been defined previously in the paragraph or chapter.

Unit identification—2/12th (slash indicates infantry), if a company (C) is identified it would read C/2/12 or C 2/12th—2 for battalion and 12 for regiment;—2-34th (dash indicates armored unit); 2 for squadron and 34 for battalion; artillery or ARTY is by battery (A, B, C, etc.), battalion (BN) (1.2.3.etc), field artillery (FA). i.e. B 1/8th FA or B/1/8 FA, B Batt, 1st BN, 8th FA. Units shown with an (M) are mechanized, mostly M113 APC's.

VC or Viet Cong—Local guerrilla militia supporting North Vietnam in combat operations.

Vietnamese language—If you want to see or pronounce more words used, visit www.212warriors.com/speak_viet.html and www.212warriors.com/rvn_gloss.html.

Weapon—A rifle, grenade launcher, or machine gun.

WIA—Wounded in action, also referred to as a "Whiskey."

World—Good ole U.S.A. aka "back in the World."

WP or Willie Pete—white phosphorus, either as a grenade or artillery round, produces an incendiary white explosion.

X.O.—Executive Officer, second in command to the company or battalion commander.

ROSTER OF CHARLIE COMPANY ENLISTED (DRAFTED) MEN

MARCH 1968 TO MARCH 1969

Rank/Grades of soldiers not updated; shows rank from records dated April and October, 1968; list may not be complete. Names in BOLD letters were Killed in Action

NAME	GRADE	MOS	DEROS	PLT
Abernathy, David	E3	11B10	Jan-70	
Abernathy, Timothy	E5	11B40	Aug-69	2nd
Abney, Wayne E Jr.	E1	11B10	Feb-69	2nd
Aguilar, Robert	E3	11B10	Feb-70	
Allen, Unkn	E3	11B10		3rd
Anderson, James A	E3	11B10	Mar-70	
Anderson, Jessie L	E3	11B10	Mar-69	1st
Anguiano, Andrew J	E3	11B10	Feb-70	
Antu, Juan	**E3**	**11B10**	**Feb-69**	**3rd**
Atkinson, Danny	E3	11B10	Nov-69	2nd
Atkinson, Nathaniel	E4	11B20	Mar-69	

NAME	GRADE	MOS	DEROS	PLT
Ausbern, John R	**E3**	**11B10**	**Sep-69**	**1st**
Bachman, George T	E3	11B10	Mar-70	
Bacon, Kent L	E3	11B10	Sep-69	1st
Badillo, Bonilla Jose	E3	11B10	Feb-69	1st
Badillo, Charles E	E2	11B10	Jul-69	
Banning, Roger L	E5	11B40	Oct-69	
Barnes, Ellis D	E3	11B10	Sep-69	3rd
Barnett, Leon	E3	11B10	Mar-69	1st
Barresi, Donald N	E5	11B40	Jul-69	
Bartolomeo, William	E4	11B20	Dec-69	
Becker, Vernon L	E4	11B20	Oct-69	3rd
Begin, Gabriel J	E3	11B10	Aug-69	
Bellamy, Frank L Jr.	E3	11B10	Jun-69	
Beltran, Robert L	**E3**	**11B10**	**Jul-69**	**3rd**
Belue, Gene F	E4	11B20	Nov-68	
Bennett, Dennis R	E5	11B40	Jun-69	
Bishop, Albert L	E2	11B10	Sep-69	
Bishop, Hugh L II	E5	11B40	Oct-68	3rd
Black, Herbert L	E4	11C20	Mar-69	4th
Blackburn, Ronnie	E3	11B10	Nov-69	1st
Bolden, Johnny J	E3	11B10	Mar-70	
Bonner, Don W	**E4**	**11B20**	**Apr-69**	**1st**
Boso, Carl R	E4	11B10	Sep-68	3rd
Bowden, Billie J	E3	11B10	Dec-69	3rd
Boyce, Argyle	E4	11B20	Oct-69	
Boyd, James O	E4	76Y20	Feb-69	
Bradley, James W	E3	11B10	Mar-70	
Brady, Daniel M	E3	11B10	Dec-69	3rd
Brasher, Willard	E4	11B10	Jul-68	3rd
Brockmann, Gerald	E3	11B10	Sep-69	1st

NAME	GRADE	MOS	DEROS	PLT
Brown, Dexane	E4	11B20	Apr-70	
Brown, Donald H	E3	11B10	Mar-70	
Brown, Raymond T	E3	11B10	Aug-69	2nd
Byrd, Gary L	E3	11B10	Aug-69	2nd
Buckley, Dennis R	E3	11B10	Aug-69	3rd
Byrum, Bobby	E5	11B40	Feb-69	1st
Cable, Walter T	E7	11B40	Aug-69	
Cain, Glenie W	E4			
Callahan, John G	E3	11B10	Sep-69	2nd
Callison, James	E4	11B20	Nov-69	
Carpenter, William	E5	11B40	Jul-69	
Carson, Clarence J	E3	11B10	Dec-69	3rd
Carter, Walter M	E3	11B10	Aug-68	
Caster, Robert J	E3	11B10	Nov-69	2nd
Champagne, Lucien	E4	11B20	Dec-69	
Chapman, William L	E3	11C10	Feb-69	4th
Chism, Freddie L	E3	11B10	Mar-69	1st
Christianson, Kenneth	E3	11B10	Apr-69	3rd
Christy, Charles S	E6	11B40	May-68	
Clardy, Patrick	E4	11B20	Dec-69	3rd
Cochran, William C	E4	11B20	Aug-68	
Coleman, Craig R	E3	11B10	Jun-69	3rd
Coleson, Richard B	E4	11B20	Sep-69	3rd
Coller, Ed	E3	11B10	Apr-69	2nd
Collins	E3	11B10	Dec-69	3rd
Conley, Alex B	**E5**	**11B40**	**Sep-69**	**1st**
Conlin, Richard J	**E5**	**11B40**	**May-69**	**3rd**
Cook, Jimmy L	E6	11B40	Apr-68	1st
Cooper, Donald	**E4**	**11B20**		
Corum, Terry L	E5	11B40	Jul-69	3rd

NAME	GRADE	MOS	DEROS	PLT
Coulter, Robert W	E5	11B40	Aug-68	2nd
Counts, Gerald D	E4	11B20	Aug-68	2nd
Crabtree, Walter A	E3	11B10	UKN	
Crank, Douglas	E3	11B10	Sep-69	4th
Crow, Elmer E Jr	E2	11B10	Dec-69	1st
Crowe, Eddie C	E3	11B10	Dec-69	
Crumrin, Gary L	E5	11B40	Sep-69	1st
Cruse, Ronald T	E4	76Y20	Jan-69	
Cuneo, Peter M	E4	11H20	Jul-68	
Dash, Unkn	E3	11B10	Dec-69	3rd
Davis, Harry T	E4	11B20	Aug-68	
Davis, John	E3	11B10	Mar-69	
Davis, Johnny V	E6	11B40	Aug-68	
Davis, Kenneth E	E3	11B10	Feb-69	1st
Deglasio, Anthony	E5	11B40	Feb-70	
Deimler, Richard	**E5**	**11B40**	**Aug-69**	**1st**
Delong, Robert L	E6	11B40	Feb-69	1st
Dewitt, Harold D	E4	11B20	Jan-69	2nd
Digiacobbe, Ron	E5	11B40	Sep-68	
Dipetrillo, Arthur	E3	11B10	Jul-69	4th
Divan, Stephen A	E3	11B10	Mar-69	1st
Dixon, James R	E5	11B40	Aug-68	
Dixon, Leamon	E4	11B20	Oct-68	3rd
Dodrill, Roger	E4	11B20	Sep-68	
Dodson, Leonard "Doc"	E5	Medic	Feb-70	
Dover, Andrew Jr.	E3	11B10	Sep-69	
Drewry, Donald	E5	11B40	Apr-70	
Duke, Darryl E	E4	11B20	Jan-69	1st
Durden, Grover D	E3	11B10	Mar-70	3rd
Duval, Alan J	E4	11B20	Sep-68	2nd

NAME	GRADE	MOS	DEROS	PLT
Dyer, Jerry	E4	11B10	Dec-69	2nd
Echols, Michael L	E3	11B10	Jul-69	2nd
Eley, Stephen W	E3	11B10	Apr-70	
Elkins, Michael B	E5	11C40	Sep-68	4th
Emeigh, Michael G	E5			
Estenes, Martin J	E4	11B10	UKN	2nd
Evers, Gordon C	E4	11C10	Jul-68	4th
Eveslage, Donald P	E3	11B10	Aug-69	
Federline, Marion "Chip"	E4	11B20	Aug-69	2nd
Ferguson, Thomas L	E5	11B40	Jul-68	1st
Findlay, Chet B	E3	11B10	Jul-69	3rd
Flood, Patrick D	E3	11B10	Aug-69	3rd
Fowler, Sidney F	E4	11B10	Mar-69	3rd
Francis, Gary L	E3	11B10	Aug-69	
Franken, Arlen D	E3	Medic		
Freidig, Dale D	E5	11B40	Jun-69	2nd
Froeschle, Gary A	E3	11B10	Mar-70	
Fullam, Harold	E5	11B40	Apr-70	
Furuta, Darol M	E4	11B10	Dec-69	1st
Garrett, Robert A	E3	11B10	Sep-68	2nd
Gillam, Clyde L Jr	E3	11B10	Sep-69	2nd
Glass, Dave	E4	11B20	Apr-68	2nd
Glasson, Charles I	E5	11B40	Sep-69	
Gleason, Robert E	E4	11B20	Feb-70	3rd
Gomez, Rudy S	E3	11B10	Jan-70	2nd
Gooding, Primus Jr	E3	11B10	Aug-69	3rd
Goodson, Larry	E3	11B10	Dec-69	2nd
Gosnell, Walter E	E3	11B10	Mar-70	3rd
Grace, Thomas M	E3	11C10	Feb-69	4th
Granum, Michael J	E5	11B40	Apr-69	1st

NAME	GRADE	MOS	DEROS	PLT
Green, Garry D "Doc"	E4	Medic	Dec-68	
Gries, Bennett A	E3	11B10	Jul-69	3rd
Grifin, Larry C	E4	11B20	Aug-68	
Gutierrez, Aleonso	E3	11B10	Mar-70	
Hahn, Michael L	E3	11C10	Feb-69	4th
Hale, Phillip A	E4	11B20	Aug-68	2nd
Hall, Curtis R	E4	11B20	Aug-68	
Hall, Dale	E3	11B10	Jan-70	3rd
Hall, Donald E	E3	11B10	Mar-69	1st
Hamilton, Carlton	E3	11B10	Apr-70	
Hampton, Larry R	E4	11B20	Sep-68	2nd
Happers, Begern	E3	11B10	Mar-70	
Harlin, Clyde A	E3	11B10	Jan-69	
Harrell, Ellis B	E6	11B40	Jan-69	
Harresi, Donald	E3	11B10	Jul-69	
Harrington, Harry	E3	11B10	Nov-69	
Harris, Craig	E4	11B20	Nov-69	
Hdekstra, Charles	E3	11B10	Aug-69	
Helmes, John E	E3	11B10	Aug-69	
Henry, Richard E	E4	11B20	Aug-68	
Herrero, Steven	E1	11B10	Aug-68	
Hickman, Everett L	E5	11B40	May-68	
Hildebrandt, Harry	E5	11B40	Jul-68	
Hill, William	E6	11B40	Mar-69	3rd
Hinricks, Ralph	E3	11B10	Oct-70	
Hobbs, Roy	E5	11B40	Nov-69	
Holt, Johnnie E	E3	11B10	Jul-69	
Holtan, Charles B	E3	11B10	Jul-69	
Hood, Arthur J	E5	11C20	Feb-69	4th

NAME	GRADE	MOS	DEROS	PLT
Hoppes, Brian	E3	11B10	Dec-69	2nd
Hornelas, Ismel F	**E3**			
Horton, James A	E4	11B20	Sep-68	
Hoskins, Alfred L	E6	11B40	Mar-70	
Houck, John	E4	11C20	Mar-69	
Hunt, Charles	E5	11B4P	Apr-70	
Hunt, Harmon	E5	11B40	Mar-69	2nd
Imbriani, Ray	E3	11B10	Apr-70	
Jackimski, Stanley	E3	11B10	Jul-69	
Jackson, Charlee	E3	11B10	Jul-69	
Jackson, Roger D	E5	11B40	May-68	
Jacobson, Jerry	E3	11B10	Sep-69	
Janney, Jackie R	E4	11C20	Feb-69	4th
Jenkins, Alvin, F	E3	11B10	Feb-70	
Jenkins, Earl H	E4	11B20	Nov-68	
Jenkins, John M	E3	11C10	Mar-70	4th
Jochum, Ronald	E3	11B10	Dec-69	3rd
Johnson, Harold R	E8	HB50	Apr-69	
Johnson, James R	E3	11C10	Feb-69	4th
Johnson, Ronnie	E5	11B40	May-68	
Johnston, Donald L	E4	11B20	Feb-70	
Jones, George W	E3	11B10	Feb-70	
Jones, Henry	E4	11B20	Jun-69	
Jones, Loren Cecil	**E4**	**11B20**		**2nd**
Judge, Roger D	E3	11B10	Aug-69	
Keatts, Lynwood C	E6	11B40	Jun-69	3rd
Kennedy, Roy E	E4	11B20	Jul-68	
Kindle, William H "Doc"	E4	Medic	Feb-70	
King, John A	E3	11B10	UKN	

NAME	GRADE	MOS	DEROS	PLT
Knapp, Tommy D	E6	11B43	Feb-69	3rd
Kneeland, Kenneth	E3	11B10	UKN	
Koon, Orion L	E4	11C20	Aug-68	4th
Korzoorfer, Edward	E3	11B10	UKN	
Kramer, Clayton L	E3	11B10	Aug-69	
Krause, Arnold W	E6	11B40	Mar-69	3rd
Kruse, Merlin, H	E5	11B40	Mar-70	
Kuhnau, Darrell	E3	11B10	Apr-69	3rd
Lajara, Luis A	E1	11B10	Apr-69	
Lambert, Dennis M	E3	11B10	Feb-70	
Lasister, Wade E "Doc"	E5	Medic	Sep-69	
Lawrence, "Doc"	E4	Medic		1st
Lawson, Benjamin	E4	11B20	Apr-70	3rd
Le Blanc, Gerald T	**E4**	**11B20**	**Apr-69**	**2nd**
Le Gault, Victor J	E8	11G5H	Aug-68	
Lee, Alvin E	E3	11B10	Aug-69	2nd
Lemew, Riley	E3	11B10	Jan-70	
Liberator, Joseph	E4	76Y30	Mar-69	4th
Lightner, Ellsworth	E5	11B40	Mar-69	1st
Lofties, Emmit	E4	11C20	Jul-68	2nd
Loving, Hebert L	E3	11B10	Mar-70	
Lowry, William A	E3	11B10	Mar-69	C.O. RTO
Lugo, Liglano D	E3	11B10	Sep-69	
Luna, Gary	E3	11B10	Jan-70	4th
Luniewicz, Jim J	E3	11B10	Apr-70	
Lyons, Virgil R	E4	11B20	Apr-69	2nd
MacDonald, Joseph	E2	11C10	Mar-69	4th
Malott, Dean C	E3	11B10	Feb-70	
Manzanares, Thomas	E3	11B10	Mar-70	

NAME	GRADE	MOS	DEROS	PLT
Marlowe, Carl F	E4	11B10	Mar-69	
Marschall, Juergen	E3	11B10	Apr-70	
Martin, Alvie E	E3	11B10	Aug-69	3rd
Martsolf, Alfred W	E5	11B40	Aug-69	
Marziano, Hiram J	E6	11B40	Mar-69	1st
Maysonet, Oscar	E3	11B10	Apr-70	
McCloud, Lawrence	E3	11B10	Apr-69	1st
McCullar, L C	E3	11B10	Mar-70	
McDermott, Clarence	E5	11B40	Jul-68	
McGary, Ralph E	E7	11C40	Mar-70	
McGeath, Richard	E3	11B10	Jul-69	3rd
McGough, Richard	E4	11B20	Jul-68	
McInvale, James D	E3	11B10	Aug-69	3rd
McIver, Duane	E4	11B20	Jan-70	
McKenzie, Donald F	**E6**	**11B40**	**Oct-69**	**1st**
McKenzie, Henry N	E3	11B10	Aug-69	
McLain, David R	E7	11B40	Apr-70	
Meyer, Bob "Doc"	E4	Medic	Dec-69	3rd
Michelson Louis L	E3	11C10	Aug-68	4th
Mikita, John	E3	11B10	Dec-69	3rd
Miller, Charles, E	E4	11B20	Apr-69	2nd
Miller, Jerry	E4	11B20	Jan-68	2nd
Miller, Percy W	E5	11B40	Jun-69	3rd
Miskimins, Floyd L	E3	11B10	Mar-69	
Molan, Jerry L	E5	11B40	UKN	4th
Monts, Edward	E3	11B10	UKN	
Moorehead, Michael	E3	11B10	Mar-70	
Morgan, Joseph O	E2	11B10	Nov-68	
Morrill, Jerry T	E5	11B40	Sep-68	2nd

NAME	GRADE	MOS	DEROS	PLT
Murphy, John M	E3	11B10	Mar-70	
Murray, Theodore	E5	71B30	Jul-68	
Nagano, Sumio	E3	11B10	Mar-69	4th
Neely, Dale W	E5	11B40	Feb-69	4th
Nelson, Earl W	E3	11B10	Apr-68	
Nevers, Gregory A	E3	11B10	Feb-69	2nd
Newhouse, Craig A	E3	11B10	Mar-70	
Noeller, Paul E	E4	11B20	Oct-68	
Nolan, Jerry D	E3	11B10	Sep-69	
Novak, Chester M	E3	11B10	Sep-69	3rd
Opplinger, Daniel J	E5	11B40	Jun-69	2nd
Ortiz-Malave, Anto	E3	11B10	Mar-69	
Orton, William G	E3	11B10	Jun-69	
Partee, John L	E6	11B4L	May-69	3rd
Pastures, Robert A	E7	11B40	Jan-69	
Pausha, Donald	E3	11B10	Dec-69	
Peda, Damian T	E3	11B10	Mar-70	
Pena, Elias R	E4	11B20	Dec-68	
Perry, Hubert M	E3	11B10	Dec-68	1st
Phelps, Daniel L	E4	11B10	Aug-68	
Phillips, Joseph	E3	11B10	Aug-69	
Phillips, Willie	E1	11B10	Aug-68	
Polus, Richard A	E3	11B10	Apr-69	3rd
Portalatin, Sigfred	E3	11B10	Mar-69	2nd
Potts, William M	E3	11B10	Mar-69	3rd
Pratt, Adrian E	E4	11B10	Nov-68	
Prevette, Gary D	E3	11B10	Sep-69	2nd
Price, Barry	E5	11B40	Mar-69	3rd
Primer, Allen W	E3	11B10	Jul-69	1st
Quesnell, Paul V	E3	11B10	Apr-69	

NAME	GRADE	MOS	DEROS	PLT
Quesnoy, Edward B	E4	11B10	Sep-68	2nd
Quintanilla, Arturo	E4	11B20	Jul-69	3rd
Raether, Lawrence	E3	11C10	Feb-69	4th
Ramirez, Ralph R	E3	11B10	Jun-68	
Ray, Ellison	E4	11B20	Jul-69	
Redner, Raymond	E3	11C10	Aug-68	4th
Reed, Bruce	**E4**	**11B20**	**Oct-69**	
Reed, Ronald L	E3	11B10	Aug-69	4th
Reitman, George A	E4	11B20	Apr-69	2nd
Ricker, John	E3	11B10	Feb-70	
Ridley, George Jr	E7	11B40	UKN	
Rinkle, Michael J	E3	11B10	Mar-69	3rd
Ritchie, Chris	E4	11B20	Feb-70	
Ritzman, Robert D	E3	11B10	Mar-69	
Roberts, Ralph, E	E3	11B10	Sep-69	
Robinson, Jessie J	E3	11B10	Mar-69	
Robinson, T.J.	E5	11B40	Mar-70	
Rodriguez-Medina, S	E3	11B10	Jan-69	
Rodriquez, Miguel	E3	11B10	Sep-69	2nd
Roehmer, Robert P	E6			2nd
Rucci, Joseph	E2	11B10	Sep-69	1st
Ryan, Lawrence	E3	11B10	Jul-69	
Sablan, Antonio	E3	11C10	Jul-68	4th
Sanson, Bernard G	E3	11C10	Apr-70	4th
Santistevan, Felix	E1	11B10	Apr-69	1st
Satterthwaite, Richard	**E3**	**11B10**	**Sep-69**	**3rd**
Scheidnes, Philip	E4	11B20	Dec-69	
Schenk, Allan D	E3	11B10	May-69	2nd
Schoonderwoerd, Craig	E6	11B40	Mar-69	1st
Schultz, David J	**E3**	**11B10**	**Mar-69**	**3rd**

NAME	GRADE	MOS	DEROS	PLT
Scogin, Richard L	E3	11B10	Mar-69	1st
Serrano, Carlos	E4	11B10	Jun-69	3rd
Sheffield, James T	**E1**	**11B10**	**Jun-69**	**1st**
Simone, James L	E3	11B10	Mar-69	1st
Sloan, James B	E6	11B40	Jan-69	
Smith, Calvin M	E5	76Y40	Sep-68	
Smith, Jerry N	E4	11B20	Sep-69	2nd
Snider, Ronald	E5	11B40	Apr-70	
Solberg Jr., Kjell	E4	11B20	Sep-68	1st
Solem, Bruce L	E4	11B20	Dec-69	3rd
Solomon, Leavy "Doc"	**E4**	**Medic**	**Sep-69**	**3rd**
Solomon, Will	E3	11B10	Aug-68	
Sovey, Thomas G	E3	11B10	Aug-69	3rd
Sparks, Ronald K	E6	11C40	Oct-69	4th
Spoores, John M	E5	11B40	Sep-68	3rd
Starling, Joe	E3		Dec-69	3rd
Steele, Harold A Jr	E7	11B40	Sep-68	
Stephens, Larry E "Doc"	**E4**			
Stepsie, Ronald S	**E3**	**11B10**	**May-69**	**3rd**
Sterling, Joe W	E4	11B20	Jul-68	
Stewart, James R	E3	11B10	Mar-70	
Stewart, William O	E1	11B10	Aug-68	2nd
Stone, Jimmy R	E3	11B10	Mar-69	
Stoney, Gerald	E4	11B20	Nov-69	
Stopa, Walter	E5	11B40	Aug-68	3rd
Studebaker, Dale C	E5	11B40	Jan-70	3rd
Sumpter, Jimmy	E3	11B10	Apr-70	
Swain, Teddy W	E5	11B40	Feb-69	
Taisler, Joseph	**E3**	**11B10**	**Apr-70**	
Taitague, John	**E4**	**11B20**	**Aug-68**	

NAME	GRADE	MOS	DEROS	PLT
Tassen, Curtis	E4	11B20	Dec-69	3rd
Tatum, Douglas L	E3	11B10	Feb-69	
Thompson, Richard	E4	11B20	Mar-69	1st
Thompson, Robert E	E4	11B20	Jul-68	
Tice, Rondual E. "Doc"	E4	Medic		
Tilley, Hubert	**E6**	**11B40**	**Apr-70**	
Tostado, Jesse	E4	11B10	Sep-69	3rd
Toto, George J	E4	11B10	Aug-69	3rd
Trahan, Ronald M	E4	11B20	Apr-69	4th
Tulupan, Polyvios	E3	11C10	Dec-68	4th
Turcotte, Greg	E3	11B10	Jan-69	3rd
Turner, Vernon	E3	11B10	Dec-69	
Tyron, Lee, Jr.	**E3**			
Tyser, William	E3	11B10	Nov-69	
Urueta, Richard	E3	11B10	Apr-70	
Valladares, Juan A	E3	11B10	Mar-69	
Varner, John E	E3	11C10	Jan-69	4th
Wadkins, Ralph L	E4	11C20	Jul-68	4th
Wahrenbrock, Andy "Doc"	E5	Medic	Jan-69	2nd
Wales, Edward A	E3	11B10	Apr-69	3rd
Walsh, John L	E3	11B10	Jun-68	2nd
Walton, Robert A	E3	11B10	Oct-69	3rd
Ward, Steve	E4	11B20	Jun-69	3rd
Warner, Arthur L	E3	11B10	May-69	3rd
West, Donald	E3	11B10	Mar-70	3rd
Whiley, Buddy	E4	11B20	Jul-69	
Whitlow, Michael D	E3	11B10	Feb-69	
Wilkes, Robert	E3	11B10	Apr-69	
William Potts	E3	11B10	Mar-69	3rd

NAME	GRADE	MOS	DEROS	PLT
Williams, John M	E3	11B10	Aug-69	
Williams, Ronald E	E5	11B40	Jul-68	
Wilson, Jerome	E4	11B20	Apr-70	
Woodson, Willie D	E3	11B10	Feb-69	2nd
Woody, Thomas	E3	11B10	Apr-70	
Woomer, Robert W	E3	11C10	Nov-68	4th
Yanez, Victor	**E3**	**11B10**	**Jan-70**	
Yanish, Joseph C	E3	11B10	Mar-69	
Yates, Tommy R	E3	11B10	Jul-69	
Ybarra, Frank	E5	11B40	Mar-69	4th
Young, Clyde W	E2	11B10	Jun-69	2nd
Young, Mevelle	E3	11B10	Mar-70	
Young, Rickie L	E4	11B20	Jul-68	
Young, Willie E	E4	11B20	Jul-69	1st
Zack, Michael E	E3	11B10	Jul-69	2nd
Zimmerman, Billy L	E4	11B20	Mar-69	1st
Zimmerman, Russell	E5	11B40	Jul-69	3rd

BATTALION COMMANDERS (FLAME 6)

LTC Rafael "Dean" Tice	Sep 67 — Mar 68
TC Charles J Bauer	Mar 68 — Apr 68
LTC Donald J Green	Apr 68 — Oct 68
LTC Thomas F Dreisenstok	Oct 68 — Feb 69
LTC John P Grisham	Feb 69 — Mar 69

CHARLIE COMPANY COMMANDERS (CHARLIE 6) AND PLT LDRS.
MARCH 1968 — MARCH 1969 CHRONOLOGICAL ORDER

1LT Jay L Hickey	Mar 68 — May 68
1LT Ronald Hendricks	Jun 68 — Aug 68
1LT Jimmy Ford	June 68 (2 weeks) Temp
CPT William N Parish	Sep 68 (week) Temp
1LT R W "Bud" McDaniel	Sep 68 — Dec 68
1LT David Riggs	Dec 68 — (2 weeks) Temp
CPT Melvin "Lee" Morrow	Jan 69 — Jun 69

PLATOON LEADERS (CHARLIE 16, 26, 36, 46)

1LT Terry Keehn — 1st Plt
1LT Jimmy Ford — 2nd Plt
1LT Hugh Vandervoort — 3rd Plt
2LT Chris Brown — 3rd Plt
1LT Laurene E Johnson
1LT Patrick J Kiggins
1LT Robert W Norris
1LT R.W. "Bud" McDaniel — 1/3rd Plt
1LT Richard Wiggins — 2nd Plt
1LT James S Hawthorne — 1st Plt
1LT James Merrett — 3rd Plt
1LT Michael Sheehan — 3rd Plt
1LT Paul Compton — 4th Plt
1LT Louis "Duke" Gietka — 1/3rd Plt
1LT David Riggs — 3rd Plt
1LT Robert Simpson
1LT Danny Mull — 3rd Plt

COMBAT KILLED BY YEAR

Co.	'66	'67	'68	'69	'70	'71	Total
HHC	0	14	06	19	04	01	44 (13%)
A	0	20	24	27	13	01	85 (26%)
B	0	20	21	22	03	01	67 (21%)
C	1	13	24	16	05	03	62 (19%)
D	0	05	21	29	0	05	62 (19%)
E	0	0	03	02	03	0	08 (02%)
TOTAL	01	72	99	115	30	11	328

THE TOP 100 TUNES — 1968

	TITLE	ARTIST(S)
1	"Hey Jude"	The Beatles
2	"Love is Blue"	Paul Mauriat
3	"Honey"	Bobby Goldsboro
4	"(Sittin' On) The Dock of the Bay"	Otis Redding
5	"People Got to Be Free"	The Rascals
6	"Sunshine of Your Love"	Cream
7	"This Guy's in Love With You"	Herb Alpert
8	"The Good, the Bad and the Ugly"	Hugo Montenegro
9	"Mrs. Robinson"	Simon & Garfunkel
10	"Tighten Up"	Archie Bell & the Drells
11	"Harper Valley PTA"	Jeannie C. Riley
12	"Little Green Apples"	O. C. Smith
13	"Mony Mony"	Tommy James and the Shondells
14	"Hello, I Love You"	The Doors
15	"Young Girl"	Gary Puckett & The Union Gap
16	"Cry Like a Baby"	The Box Tops
17	"Stoned Soul Picnic"	The 5th Dimension
18	"Grazing in the Grass"	Hugh Masekela
19	"Midnight Confessions"	The Grass Roots
20	"Dance to the Music"	Sly & the Family Stone

21	"The Horse"	Cliff Nobles
22	"I Wish It Would Rain"	The Temptations
23	"La-La (Means I Love You)"	The Delfonics
24	"Turn Around, Look at Me"	The Vogues
25	"Judy in Disguise (With Glasses)"	John Fred & His Playboy Band
26	"Spooky"	Classics IV
27	"Love Child"	The Supremes
28	"Angel of the Morning"	Merrilee Rush
29	"The Ballad of Bonnie and Clyde"	Georgie Fame
30	"Those Were the Days"	Mary Hopkin
31	"Born to Be Wild"	Steppenwolf
32	"Cowboys to Girls"	The Intruders
33	"Simon Says"	1910 Fruitgum Company
34	"Lady Willpower"	Gary Puckett & The Union Gap
35	"A Beautiful Morning"	The Rascals
36	"The Look of Love"	Sérgio Mendes
37	"Hold Me Tight"	Johnny Nash
38	"Yummy Yummy Yummy"	Ohio Express
39	"Fire"	The Crazy World of Arthur Brown
40	"Love Is All Around"	The Troggs
41	"Playboy"	Gene & Debbe
42	"(Theme from) Valley of the Dolls"	Dionne Warwick
43	"Classical Gas"	Mason Williams
44	"Slip Away"	Clarence Carter
45	"Girl Watcher"	The O'Kaysions
46	"(Sweet Sweet Baby) Since You've Been Gone"	Aretha Franklin
47	"Green Tambourine"	The Lemon Pipers
48	"1, 2, 3, Red Light"	1910 Fruitgum Company
49	"Reach out of the Darkness"	Friend & Lover
50	"Jumpin' Jack Flash"	The Rolling Stones

51	"MacArthur Park"	Richard Harris
52	"Light My Fire"	José Feliciano
53	"I Love You"	People!
54	"Take Time to Know Her"	Percy Sledge
55	"Pictures of Matchstick Men"	Status Quo
56	"Summertime Blues"	Blue Cheer
57	"Ain't Nothing Like the Real Thing"	Marvin Gaye & Tammi Terrell
58	"I Got the Feelin'"	James Brown
59	"I've Gotta Get a Message to You"	The Bee Gees
60	"Lady Madonna"	The Beatles
61	"Hurdy Gurdy Man"	Donovan
62	"Magic Carpet Ride"	Steppenwolf
63	"Bottle of Wine"	The Fireballs
64	"Stay in My Corner"	The Dells
65	"Soul Serenade"	Willie Mitchell
66	"Delilah"	Tom Jones
67	"Nobody but Me"	The Human Beinz
68	"I Thank You"	Sam & Dave
69	"The Fool on the Hill"	Sérgio Mendes
70	"Sky Pilot"	The Animals
71	"Indian Lake"	The Cowsills
72	"I Wonder What She's Doing Tonight"	Tommy Boyce & Bobby Hart
73	"Over You"	Gary Puckett & The Union Gap
74	"Goin' Out of My Head"	The Lettermen
75	"Shoo-Be-Doo-Be-Doo-Da-Day"	Stevie Wonder
76	"The Unicorn"	The Irish Rovers
77	"You Keep Me Hangin' On"	Vanilla Fudge
78	"Revolution"	The Beatles
79	"Woman, Woman"	Gary Puckett & The Union Gap
80	"Elenore"	The Turtles

81	"White Room"	Cream
82	"You're All I Need to Get By"	Marvin Gaye & Tammi Terrell
83	"Baby, Now That I've Found You"	The Foundations
84	"Sweet Inspiration"	The Sweet Inspirations
85	"If You Can Want"	Smokey Robinson and the Miracles
86	"Cab Driver"	The Mills Brothers
87	"Time Has Come Today"	The Chambers Brothers
88	"Do You Know the Way to San Jose"	Dionne Warwick
89	"Scarborough Fair"	Simon & Garfunkel
90	"Say It Loud—I'm Black and I'm Proud"	James Brown
91	"The Mighty Quinn"	Manfred Mann
92	"Here Comes the Judge"	Shorty Long
93	"I Say a Little Prayer"	Aretha Franklin
94	"Think"	Aretha Franklin
95	"Sealed with a Kiss"	Gary Lewis and the Playboys
96	"Piece of My Heart"	Big Brother and the Holding Company
97	"Suzie Q."	Creedence Clearwater Revival
98	"Bend Me, Shape Me"	The American Breed
99	"Hey, Western Union Man"	Jerry Butler
100	"Never Give You Up"	Jerry Butler

HALF TRUTHS AND LIES

Stories you will hear...

Newbies and replacements were always put out on point or flank and the veterans stayed in the columns—50/50 chance—a smart sergeant would not put an newbie on point because he didn't know what to look for or what to do. Being on flank as the second flanker trailing a vet was most likely until he gained experience

I walked point every day—False, only if your platoon had point duties which it did not; it was a rotational duty

We went out on ambush every night—False, normally it was rotational between platoons or companies

I was a tunnel rat—only if you were the smallest guy in the platoon, otherwise you couldn't get down in the tunnel

I put notches on my M16 for all the kill's I had...You be the judge. Not everyone carried a knife and the stock of an M16 was a pretty tough composite material

I set a record for the most trips to the First Aid station—True if getting cured of the "CLAP" or you suffered from Dysentery—About 70% earned one Purple Heart

I was a sniper in my company—False—LRRP or RECON or Special Forces, yes,—in a line company like the 2/12th, NO.

I was somewhere in Cambodia –False, unless it was in June, 1970 when we invaded that country or if you were Special Ops.

I extended my tour by three months so I could exit the Army when I got home or to prevent a relative from serving at the same time—True

Everyone spoke Vietnamese—False, maybe a few words unless you were sent to language school in Monterey, California.

I liked the AK-47 over the M16—only if you've never fired an M16

Officers and RTO's were targeted to be killed first—Not true, the number who died in the line of duty does not exceed the deaths of anyone else. Out of 328 KIA's in my battalion, only 2 company commanders and 14 platoon leaders were killed and a few of these were killed by mortar fire, so they were not directly targeted. That represents 5% of the total number. I only have knowledge of one RTO being killed, but there could be more.

ABOUT THE AUTHOR

Drafted in the U.S. Army in September 1967 and left the military in September 1969

I arrived in Vietnam as a Private First Class, and became a member of Charlie Company, Third Platoon, 2nd Battalion, 12th Infantry Regiment of the 25th Infantry Division. I served as an ammo bearer on a M60 machine gun crew, then became an RTO for the platoon leader, was promoted to SGT in August, and took over 3rd rifle squad. In December, I acted as PSG for a month, was promoted to SSG in February before returning home. I was awarded the Bronze Star, Purple Heart w/OLC, CIB and other awards.

In 2011 I launched a website, www.212warriors.com dedicated to the unit's time in Vietnam. Somehow, in 2013, the 2/12th Inf which is currently active, discovered this website. The U.S. Army is very dedicated to keeping active contact with former veterans. Because of this practice, they extended an invitation to me and other veterans I could convenience, to go to Fort Carson, Colorado, to attend a "Dining-In" event. The Dining-In Event is for the battalion and attended by all officers and enlisted wearing their dress blue uniforms, about 650 strong. I had the honor of being the guest speaker. Joining me at this event was Bill Braniff, A Co, Craig Schoonderwoerd and Steve Ward from C Co. I have traveled there on numerous occasions to interact with the present commanding officers of the unit, their CSM's and others at change of command ceremonies and hosted events. These trips are currently ongoing as of this writing in 2019. I was also on the executive committee headed by LTC Ed Northrop (commander of Charlie Co, 1/12th, 1967, as a Captain), the current honorary regimental colonel, in raising funds to erect our own monument for the 12th Infantry Regiment at the Walk of Honor, near the parade field at the National Infantry Museum, Fort Benning, Georgia.

Following the dedication ceremony on May 31, 2017, by Order of the Secretary of the Army, in recognition of outstanding contributions to Regimental continuity, committee members were granted and assigned the distinction of Distinguished Member of the 12th Infantry Regiment (DMOR) honors.

www.ingramcontent.com/pod-product-compliance
Lightning Source LLC
Chambersburg PA
CBHW030145100526
44592CB00009B/121